Lecture Notes in Computer Science 9380

Commenced Publication in 1973
Founding and Former Series Editors:
Gerhard Goos, Juris Hartmanis, and Jan van Leeuwen

More information about this series at http://www.springer.com/series/8851

Ngoc Thanh Nguyen · Ryszard Kowalczyk
Fatos Xhafa (Eds.)

Transactions on
Computational
Collective Intelligence XIX

 Springer

Editors-in-Chief
Ngoc Thanh Nguyen
Department of Information Systems
Wrocław University of Technology
Wrocław
Poland

Ryszard Kowalczyk
Swinburne University of Technology
Hawthorn, VIC
Australia

Guest Editor
Fatos Xhafa
Universitat Politècnica de Catalunya
Barcelona
Spain

ISSN 0302-9743   ISSN 1611-3349 (electronic)
Lecture Notes in Computer Science
ISBN 978-3-662-49016-7   ISBN 978-3-662-49017-4 (eBook)
DOI 10.1007/978-3-662-49017-4

Library of Congress Control Number: 2015955870

Springer Heidelberg New York Dordrecht London

Springer-Verlag GmbH Berlin Heidelberg is part of Springer Science+Business Media
(www.springer.com)

Preface

It is my pleasure to present the XIX volume of LNCS Transactions on Computational Collective Intelligence (TCCI). This volume includes 11 interesting and original papers, which have been selected after a peer-review process. The papers present interesting research findings and they identify and discuss issues and challenges in the field of building collective intelligence according to new computing trends from networking and cloud computing. Specifically, the papers cover research topics such as new methodologies for information management, recommendation systems, mining and machine learning for big data, security and privacy issues, security and encryption in cloud data centers, social networks analysis, data integration, and image processing.

The papers of the special issue are arranged as follows.

The first paper, "Management and Computer Science Synergies: A Theoretical Framework for Context-Sensitive Simulation Environment" by De Maio et al., presents a theoretical framework to address contextual decision making concerning relations between commitment, loyalty, and customer satisfaction. The authors' approach is based on a full-mode generation of knowledge starting from the hypothetical assumptions relative to simulation using context data.

The second paper, entitled "Improved Recommendation System Using Friend Relationship in SNS" by Liao et al., investigates how to take better advantage of the friend relationship in SNS through improved recommendation methods and analyzes the case of Chinese users in SNS having massive users and a potential commercial value.

In the third paper, "Bidirectional Analysis Method of Static XSS Defect Detection Technique Based on Database Query Language," Cui et al. analyze the vulnerabilities of Web applications, such as XSS defects. Their method is based on database query language techniques to build a static analysis method of XSS defect detection of Java Web application by analyzing data flow reversely. The proposed technique is proven useful for big data analysis.

In the fourth paper, "A Multilevel Security Model for Search Engine over Integrated Data," Zhao et al. study issues arising in integrating multiple sources of data, such as security issue, where different sources of data may have different access control policies. Therefore the authors propose a model to integrate multiple security policies while data are integrated to ensure all data access respects the original data access control policies.

The fifth paper, entitled "Security Analysis of Two Identity-Based Proxy Re-encryption Schemes in Multi-user Networks" by Wang et al., concerns proxy re-encryption, which is to securely enable the re-encryption of ciphertexts from one key to another, without relying on trusted parties. The proxy re-encryption and its variants are very useful nowadays in the context of cloud computing. The authors give formal models for such multi-user schemes and analyze the weakness of two existing security schemes.

In the sixth paper, "Enabling Vehicular Data with Distributed Machine Learning," Chilipirea et al. address the issues arising in analyzing big data sets with data-mining and machine-learning methods. The authors show the limitations of current approaches and advocate the use of advanced parallel processing methods, models, and cloud computing infrastructures to efficiently analyze big vehicular data, a type of big data arising in vehicular networks. The approach is exemplified for the case of a k-nearest neighbors algorithm.

The next paper is titled "Adapting Distributed Evolutionary Algorithms to Heterogeneous Hardware" authored by Salto and Alba, who analyze the impact of heterogeneity in the performance of a parallel metaheuristics and their efficiency in time when executed in heterogeneous clusters. Therefore, the authors provide a methodology that enables a distributed genetic algorithm to be customized for higher efficiency on any available hardware resources with different computing power, all of them collaborating to solve the same problem.

Wang et al., in the eighth paper entitled "Eroca: A Framework for Efficiently Recovering Outsourced Ciphertexts for Autonomous Vehicles," address the question of how to ensure the security and privacy of autonomous vehicle systems by encrypting the real-time traffic information and upload the ciphertexts to the center cloud for easily sharing road traffic information among the vehicles. Therefore the authors propose a method to retrieve and update the data from the early-encrypted file in the cloud efficiently using the notion of attribute-based encryption with recoverable sender.

In the ninth paper, "Coarser-Grained Multi-user Searchable Encryption in Hybrid Cloud," Liu et al. present a new concept of coarse-grained access control and use it to construct a multi-user searchable encryption model in hybrid cloud. The authors implemented such a practical scheme using an improved searchable symmetric encryption scheme, and the obtained results support the claims of their scheme on the security analysis.

The tenth paper, "Quantum Information Splitting Based on Entangled States" by Tan et al., proposes two quantum information splitting protocols that take full advantage of the entanglement properties of Bell states and cluster states in different bases to check eavesdropping. Such protocols are proven efficiently secure against the intercept and resend attack and entangled ancilla particles attack. These protocols are of particular importance for cloud computing systems.

In the last paper, "Zero-tree Wavelet Algorithm Joint with Huffman Encoding for Image Compression," Zhang et al. employ embedded zero-tree wavelet (EZW) as an effective image encoding algorithm, by improving it through the use of zero-tree structure, wavelet coefficient scanning mode, and embedding EZQ algorithm flow. The improved algorithm is shown to not only increase the compression ratio and encoding efficient, but also improve the peak signal-to-noise ratio of images and make the vision clearer in a more feasible and effective way.

I would like to sincerely thank the authors for their valuable contributions and the reviewers for their time and constructive feedback that greatly helped authors to improve their manuscripts. I would like to thank the Editor-in-Chief of TCCI, Prof. Ngoc Thanh Nguyen, for the opportunity to edit this special issue and for his encouragement. The support by Dr. Bernadetta Maleszka, Assistant Editor of TCCI, is highly appreciated.

This work has been partially supported by funds from the Spanish Ministry for Economy and Competitiveness (MINECO) and the European Union (FEDER funds) under grant COMMAS (ref. TIN2013-46181-C2-1-R).

September 2015   Fatos Xhafa

Transactions on Computational Collective Intelligence

This Springer journal focuses on research in applications of the computer-based methods of computational collective intelligence (CCI) and their applications in a wide range of fields such as the Semantic Web, social networks, and multi-agent systems. It aims to provide a forum for the presentation of scientific research and technological achievements accomplished by the international community.

The topics addressed by this journal include all solutions of real-life problems for which it is necessary to use CCI technologies to achieve effective results. The emphasis of the papers published is on novel and original research and technological advancements. Special features on specific topics are welcome.

Contents

Management and Computer Science Synergies: A Theoretical Framework for Context Sensitive Simulation Environment

Carmen De Maio[1], Giuseppe Fenza[1], Vincenzo Loia[1]([✉]), Aurelio Tommasetti[2], Orlando Troisi[2], and Massimiliano Vesci[2]

[1] Deparment of Computer Science, University of Salerno, Fisciano, SA, Italy
{cdemaio,gfenza,loia}@unisa.it
[2] Department of Management and Information Technology, University of Salerno, Fisciano, SA, Italy
{tommasetti,otroisi,mvesci}@unisa.it

Abstract. In the light of contemporary management trends and on the basis of the theory of "open innovation", derives the concept of "crossfertilization". Crossfertilization, i.e. profitable inter-group knowledge exchange facilitates the fusion of input from different disciplinary communities. In such a scenario, the study highlights the opportunities deriving from the cross-fertilization between management and computer science domains, and yields in terms of cognitive synergies results that exceeds by far the individual outputs of the parties involved. The approach we propose is a full mode generation of knowledge starting from the hypothetical assumptions relative to simulation using context data. A general workflow complementary Structural Equation Modeling (SEM) is defining being the most appropriate mathematical technique for testing causal relationships between latent variables with Fuzzy Data Analysis techniques in order to tailor Decision Support System (DSS) to the context of application. The main contribution of our study is the definition of a theoretical framework to address contextual decision making concerning relations between commitment, loyalty and customer satisfaction.

Keywords: Computer science · Management · Structural equation modeling · Fuzzy theory · DSS

1 Introduction

In the context of today's socio-economic world, characterized by growing dynamism and eccessive competition, any organization (from private, to public sector and non-profit firms) is obliged to seek lasting sustainable competitive advantage based on unique know-how. According to contemporary management literature, ever more focused on the concepts of knowledge management and innovation, and on the basis of the theory of "open innovation" [1], the concept of "cross-fertilization", i.e. the profitable exchange of knowledge between working groups of individuals from different disciplines, fosters cutting edge ideas of

© Springer-Verlag Berlin Heidelberg 2015
N.T. Nguyen et al. (Eds.): Transactions on CCI XIX, LNCS 9380, pp. 1–16, 2015.
DOI: 10.1007/978-3-662-49017-4_1

radical innovation [2] by virtue of the fusion of input from different disciplinary communities. The outcome of the exchange of systemic interdisciplinary skills within an organization usually results in the generating of innovative technologies and enhanced products and services and a radical change in the organizational mindset.

The pervasive contribution that ICT offers the value chain *tout court* implies ongoing cross-fertilization between management, computer science, mathematics, engineering and networking [3,4]. Conseguently, from transversal competences, the sum of cognitive synergies far exceeds that of the separate outputs of the parties involved.

Therefore, widespread cooperation between individual systems of knowledge potentially means long term regeneration of both the economy and the social, through the sharing of expertise and wider participation in business processes.

The purpose of out study is to highlight the opportunities and synergies that exist between the domains of management and computer science. In particular, a full mode generation of knowledge is proposed starting from the hypothetical assumptions relative to simulation using context data. The main contribution of the work is the definition of a general workflow enabling simulation environment to perform what-if analysis enabling impact tuning relative to specific management choices (e.g., spending review, revocation of a service,etc.). The proof of concept will be tailored on a specific management case study regarding relationships between commitment, loyalty and customer satisfaction. In literature, many studies analyzing the antecedents of satisfaction and loyalty are present [5]. Usually, Structural Equation Modeling (SEM) is widely applied in these studies to predict endogenous latent variables and to confirm relationships between customer satisfaction and loyalty (causal analysis or path analysis). In particular, SEM allows to accept or reject hypothesis relative to positive/negative correlations among the variables investigated within a specific context, starting from customer opinions gathered through a questionnaire eliciting specific items. Notwithstanding, surveys may contain questions involving opinions, habits, or individuals attitudes and are often subjective and vague. Furthermore, the results of SEM are not always exploited to further perform what-if analysis. In order to mitigate such limitations, we introduce well-founded Fuzzy Data Analysis techniques to manage ambiguity and to train fuzzy classifier tailored on responses to the questionnaire. In this sense, our study argues that exploiting the causal relationships resulting from SEM it is possible to drive Fuzzy Data Analysis in order to build a simulation environment that leverages on fuzzy classifiers to support management needs and contextual decision making.

For this purpose the construction of a theoretical framework of context management (see. Sect. 2), will be simulated starting from bibliographic review (see Sect. 2.1) to identify the research model and hypotheses (see Sect. 2.2). Structural Equation Model (see Sect. 3) the most appropriate and widespread statistical- mathematical technique in management for testing the causal relationships between variables (path analysis), fuzzy techniques and DSS (see Sect. 4) will be used as a functional tool to simulate in varying contexts changes in weights load and the relations between the variables. Subsequently, the implications of potential and conducted simulations for management are reviewed (see Sect. 5). The paper concludes presenting an overview of related works and the summing up of the main contributions of this work.

2   The Proposition of a Theoretical Model

A simulated management framework designed to validate the existence of a specific relationship between commitment, loyalty and customer satisfaction will be constructed. The preliminary purpose is the illustration of a relative demo and subsequently, an analysis of the most appropriate computational intelligence techniques for simulation in different contexts. Normally the proposition of a framework in management and in science generally, begins with an in-depth review of the literature to ascertain the state of the art respect to the topic of interest to derive an adequate research hypothesis.

2.1   A Review of the Literature

Commitment. Interesting studies on commitment in the organizational and psychological field [6–8], have been analyzed in various disciplines (including management) and applied to different sectors [9–15]. The theory of commitment has extremely broad scope and studies on human commitment in the public sector [16,17] are also quite frequent. Conversely, less explored is the potential of the theory in strategic and competitive decisionmaking applied to public organizations. Few works apply the theory of commitment to public companies that provide services in non-market scenarios, as for example, universities [18].

It would be interesting therefore to study in depth the relation between student satisfaction, commitment and decisionmaking in terms of loyalty towards the university.

Relevant for the influence it had on subsequent studies was the work of [6]. In their view, the approach to psychological commitment can be divided into three components:

- *affective commitment*, i.e. emotional attachment, the sense of identification and involvement in the organization;
- *continuous commitment*, which is identified with the costs associated to leaving the organization;
- *normative commitment* or the feeling of obligation towards the organization.

As in the original model, therefore, although commitment was considered a three-dimensional construct, consumer commitment was often considered a unidimensional [19,20], or two-dimensional construct. In particular [21], found some overlapping between continuous and normative commitment. In their study of inter-firm relationships in the distribution channel, the authors identified two types of commitment:

– *affective commitment*, relating to the emotional sphere, and expresses the extent to which the members of the channel find pleasure in maintaining their relationship with specific partners;
– *calculative commitment*, to measure how members perceive the need to maintain that relationship. It strictly adheres to the sphere of costs and benefits and an assessment of the investment made in building the relationship with the partner and therefore the costs arising from potential change.

Customer Satisfaction (CS). CS is a complex construct and various definitions can be found in the literature [22]. One of the most common models for assessing CS is that proposed by [23], whereby customer satisfaction is related to consumer expectations. [24] for example, found that satisfaction is a function of the cost benefit ratio, from which it follows that price, benefits and consumer activity are all factors to take into account when analyzing CS. Furthermore, other researchers have also found it feasible to include CS in measuring consumer judgment of a product or service in terms of producing adequate levels of satisfaction relative to client needs, desires and goals [24]. [25–27] have also defined CS in similar ways.

Loyalty. Three approaches can be applied to measuring customer loyalty: the attitudinal,the behavioral and a blended approach [28]. The attitudinal perspective attempts to estimate the extent to which a consumer is favorably disposed toward a service. This is reflected in actions such as purchase recommendation of a particular product or service to other customers (word of mouth - WOM) or the commitment to give positive opinions on the service experienced [29]. The behavioral perspective is most persuasive, it observes repeated Purchase (RP) patterns: RP is measured by studying the history of each customer's purchase and repurchase. The blended or mixed approach, instead, combines the purely numerical aspect of the purchase with the intention of repurchase. This approach would appear more suitable for measuring the behavior of loyal consumers [30]. The study considers "loyal" customers, those who hold favorable attitudes towards the organization, sponsoring or recommending it to other customers and with a repurchasing behavior commitment towards the organization.

2.2   The Model and Research Hypotheses

As the importance of commitment in the development of relations between organizations with reference to the buying behavior of the consumer was highlighted, the relationship between commitment and customer satisfaction [12,31] was also tested. A further hypothesis of investigation was considered, i.e. that a model to study the extent to which university students are "committed" to the organization that provides the service training could be devised in order to evaluate whether and how the components of commitment are related to student loyalty or how these are involved in the choices that are made when selecting a university and/or a degree course (Fig. 1).

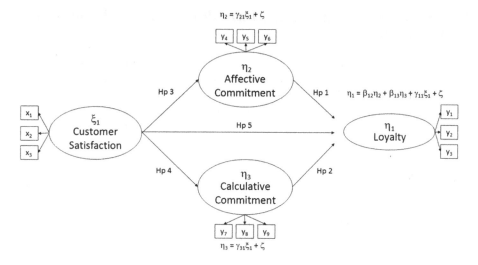

Fig. 1. The hypothetical theoretical model and research hypotheses

In literature [32] it was observed that at a high affective commitment or a high sense of identification and involvement in the organization corresponds to a high degree of loyalty **(Hp 1)**.

[33] as regards calculative commitment have stressed that the link of calculative commitment and normative commitment with loyalty is sometimes less strong than that of affective commitment.

A second research hypothesis of our framework derives:

Hp 2: Calculative commitment is positively correlated with the student's loyalty to the organization that delivers the training service.

[34], in a study on customer orientation of the employees of the organization with particular reference to the service sector, showed that affective commitment is significantly influenced by customer satisfaction. And [35] has shown that especially in the services sector, satisfaction has a directly positive effect on commitment. On the basis of the above, two additional research hypothesis of the model derive:

Hp 3: Customer satisfaction is positively correlated with affective commitment.

Hp 4: Customer satisfaction is positively correlated with calculative commitment

The link between customer satisfaction and loyalty has been repeatedly confirmed in several studies in the field of services [36–38]. It is therefore possible to posit the latest fundamental research hypothesis:

Hp 5: Customer satisfaction is positively associated with loyalty.

3 Hypotheses Validation Methodology of the Proposed Model: SEM

The framework proposed in Fig. 1 is characterized by the presence of multiple relationships between latent constructs. In this model, the dependent variable is represented by the loyalty of students to the organization and the independent variable by customer satisfaction. Commitment, in its two components, affective and calculative, is the element of mediation between satisfaction and loyalty.

Especially when researchers "deal with relations between constructs such as satisfaction, role ambiguity, or attitude, SEM is likely to be the methodology of choice" [39]. Consequently, the technique that has taken on a prominent role in managerial studies, to validate the research hypothesis among latent constructs is that of a structural equation model [40, 41] seeing as it enables the measuring and specifying simultaneously, multiple causal relationships between a set of (latent) not observed variables (commonly defined constructs) from specific observed indicators. More specifically, indicators are the items proposed in the questionnaire relative to customer opinions. On the basis of an accurate literature review and of a previous study using similar items/indicator, the indicators are utilized to derive each latent construct. As is well known, SEM is a common mathematical-statistical technique to evaluate the structure and the links of latent variables in a model of causal relationships.

The structural equation model is in effect, the only one capable of providing a representation, albeit simplified, of real processes. It takes into account the multiplicity of causes that act on a dependent variable (multivariate analysis), but also the connections between different causes.

In a more general formulation, the model can be represented as follows:

$$X_1 = b_{12}X_2 + b_{13}X_3 + \cdots + b_{1k}X_k + e_1$$
$$X_2 = b_{21}X_1 + b_{23}X_3 + \cdots + b_{2k}X_k + e_2$$
$$\cdots$$
$$X_k = b_{k1}X_1 + b_{k2}X_2 + \cdots + b_{k,k-1}X_{k-1} + e_k$$

$$(1)$$

Each equations expresses the causal link between a dependent variable, to the left of the equal sign, and a number of other variables; these variables express what variables the coefficients b depends on and in what way. The equations will be many as there are dependent variables.

Normally, the variables are indicated by ξ e η and the parameters by γ e β and error ϵ by ς, as shown in Fig. 1 for each dependent variable.

Real processes are a complex network of interactions and the approach using several equations enables the defining of the network structure from which the definition of structural equation modeling derives. The single equation of the system is therefore a "structural equation" and the coefficients b "structural parameters".

The overall pattern of the model is composed of:

- measurement model with which to describe the measurement variables of the latent variables and the type of measurement model;

– causal model (or path analysis) that describes the relationships between latent variables.

The structural equation model enables especially, the carrying out simultaneously of both a confirmatory factor analysis and a path analysis. Confirmatory factor analysis enables the evaluation of the validity of the measurement model and path analysis enables the estimation of the parameters of the causal relationship between the constructs. Precisely because the structural equation model supports simultaneous processing relationships between multiple non-observed dependent and independent variables [42], it is especially appropriate for analyzing data in the field of social research, particularly in studies on behavior and marketing [43].

Adopting specific techniques and specific software is possible, therefore, on the basis of indicators detected in the real world, to confirm the hypothesis of relationships among the latent constructs placed (not measured). The indicators are detected on samples of the population through questionnaires in which judgment about the perception of specific aspects is normally asked for. The indicators are usually defined "items" and are represented by the letters "x" and "y" as shown in Fig. 1.

The hypothesis simulated here is that we have four latent constructs (affective commitment, calculative commitment, loyalty and customer satisfaction) measured by nine indicators (three indicators, which for brevity are not illustrated here, for each latent construct), 1 (construct) independent variable (customer satisfaction) and 3 variables (constructs) customers.

It was, however, noted that "the results of SEM analysis were too vague for decision makers to identify the practical control solution". SEM is not performed well enough to clarify what-if scenario such as "What happens to purchasing intention if we drop security or privacy from causal model?" [44].

The basic idea of this study is that is possible to use the initial indicators and the results of the model of causal relations arising from the SEM in terms of confirmation/non-confirmation of hypotheses of relationships posed, particularly in computational intelligence techniques in order to offer a contribution to management in terms of strategic choices and optimal policies.

4   Methodology to Build Context Sensitive Simulation Environment: Fuzzy Data Analysis

As introduced in the previous sections, Fuzzy Data Analysis techniques are introduced in the proposed theoretical framework to build simulation environment that is automatically tailored on the collected customer opinions. Consequently, this phase is aimed at quantifying and qualifying the correlations assessed by SEM, analyzing opinions collected from interviewing people with a specific culture, located in a specific region, etc. Methodology leverage on the fuzzy mathematical model to extract fuzzy rule based classifier trained on collected opinions. Classifier enable us to estimate change in customer perceptions corresponding to a specific management decision, approximating distribution of the collected opinions between different identified levels of *Loyalty* or *Satisfaction*. The crucial steps that we identify to build fuzzy rule based classifier are essentially fuzzy clustering and fuzzy rules extraction. The former groups together similar opinions with respect to the indicators and hypothesis confirmed by SEM. The latter is aimed to identify fuzzy IF-THEN rules generated interpreting clustering results. In fact, each cluster intuitively identify the conditions (e.g., *Affective or Calculative Commitment*, etc.) determining specific level (e.g., low, medium, and so on) of selected variables (e.g., *Loyalty*). For instance, the rules in the selected case study will indicate which range of *Affective Commitment* determines specific level of *Customer Satisfaction* qualifying and quantifying positive correlation (assumed to be confirmed by SEM) between these factors that could be exploited at simulation time.

It is worthy of note that phase of *Fuzzy Data Analysis* is driven by the results of SEM. A different portion of the detected opinions will be analyzed in order to train fuzzy classifier on a confirmed hypothesis. In more detail, according to the proposed aim the theoretical framework envisages the following remits:

– *(Fuzzy) Clustering procedure*: once the customers' opinions are collected and arranged adequately, they are processed by means of clustering techniques. Data partitioning resulting by clustering execution is useful to analyze segment of population with similar opinions, but also correlations between values of the feature in each cluster. The clustering algorithm will be executed for each hypothesis selected in the investigated case study (see Sect. 2.2);
– *Fuzzy Rules Extraction*: finally, taking into fuzzy data partitioning resulting from previous phase, fuzzy rules will be extracted. Fuzzy rules will be exploited to classify opinions without clustering algorithm again seeing as they are a time consuming procedure. In this work, fuzzy rules will be injected into the simulation environment to enable an indicators (e.g., *Loyalty*) corresponding to changes, due to management choices, on the antecedent variables (i.e., *Affective Commitment, Calculative Commitment*). This phase will be carried proportionally to that of clustering algorithms, in fact, for each selected hypothesis (see Sect. 2.2) fuzzy rules set will be extracted by the proposed workflow.

The step described above are performed in a purely sequential manner and could be executed starting from the formalization of hypothesis. In any event, feedback is possible when further specifications and requirements force the revision and/or re-adaptation of one or more phases in sequence.

The following sections provide theoretical details about well-known fuzzy clustering and fuzzy rules extraction methods on which the proposed theoretical framework leverages.

4.1   Fuzzy Clustering

Well-known Fuzzy C-Means (FCM) [45] is clustering algorithm often used for customer segmentation [46]. It takes as input customers' opinions matrix and attemps to obtain an "optimal" partitioning and centroid of each retrieved partition. Each opinion is a row of matrix and each column is a student's answer in the questionnaire. FCM recognizes spherical "clouds of points" (clusters of customers' opinions) in a multi-dimensional data space (i.e., data matrix), and each cluster is represented by its center point (prototype or centroid). Specifically, the function that FCM minimizes is the weighted sum of the distances between data points $\underline{x}$ and the centroid $\underline{v}$, according to this formula:

$$V(U) = \sum_{i=1}^{c} \sum_{j=1}^{n} u_{ij}^{m} \left\| \underline{x_j} - \underline{v_i} \right\|^{2} \tag{2}$$

where $c \geq 2$ is the number of clusters, $u_{i,j} \in [0,1]$ is the membership degree of x_j in the $i - th$ cluster and m 1 controls the quantity of fuzziness in the classification process (usually in our application $m = 2$). In this approach, each row of data matrix is a vector representing the opinions (viz. the constructs identified previously) for each customer opinion $x = (x_1, x_2, x_3, x_4)$. After clustering, data partitions are returned, in a prior fixed number c of clusters. Nevertheless, Fuzzy C-Means considers a prior fixed number of clusters. In order to give more accurate partitions of data, many researchers have studied criteria of cluster validity [47]. In particular this work adopts Xie-Beni [47] index to measure the differences between clusters and to determine which of the detected changes are significant [46] in order to carry out valid partition.

4.2 Fuzzy Rules Extraction

The approach exploited to extract fuzzy rules is based on the cylindrical extension technique [48] (a projection-based method of n-dimensional argument vector). More specifically, the generic $i - th$ fuzzy cluster can be described by a fuzzy rule, where each features is described through a fuzzy set. Each fuzzy cluster K_i with $i = 1, \ldots c$ can be represented through n functions $A_{i1}, A_{i2}, \ldots, A_{in}$, obtained by the projected and interpolated memberships. Generally, the linear interpolation of memberships is computed; in order to assign linguistic label, the membership function is computed by linear interpolation of the projected membership with each axis (according to the n-dimension space), per resulting cluster. A fuzzy rule, in the form of Mamdani model [49], is described as follows:

$$R_i : \textbf{If } (x_1 \text{ } is \text{ } A_{i1})AND(x_2 \text{ } is \text{ } A_{i2})AND \ldots AND(x_{n-1} \text{ } is \text{ } A_{in-1}) \tag{3}$$

$$\textbf{then } (x_n \text{ } is \text{ } A_{in})$$

where R_i is the $i - th$ rule, $x_1, x_2, \ldots, x_{n-1}$ are the input variables, $A_{i1}, A_{i2}, \ldots, A_{in-1}$ are the membership function assigned to corresponding input variables, variable x_n represents the value of the $i - th$ rule output and A_{in} is the membership function assigned to corresponding output variable. The membership function A_{ij} is given by the formula as:

$$A_{ij}(x_j) = exp \left\{ -\frac{1}{2} \left(\frac{x_j - x_{ij}^{*}}{\sigma_{ij}} \right)^{2} \right\} \tag{4}$$

where

$$\sigma_{ij}^2 = \frac{1}{2\alpha}, \quad \alpha = \frac{4}{r_a^2} \tag{5}$$

where x_{ij}^* is the $j - th$ element of x_i^*, and r_a is a positive constant. The degree of fulfillment of each rule is defined as:

$$\mu_i = exp\left(-\alpha \|x - x_i^*\|^2\right). \tag{6}$$

5   The Simulation Environment

The Simulation Environment is aimed to support decision makers in order to evaluate some management choices and mitigate impact in terms of opinions distribution with respect to the selected hypothesis. The selected case study deals with customer opinions in terms of loyalty and satisfaction. The simulation environment resulting from the designed workflow is able to evaluate the initial situation in terms of how many customers show certain level of loyalty and/or satisfaction. It is obtained by classifying opinions by means of extracted fuzzy rules, identifying how many opinions fall within a specific range of *Loyalty* (e.g., low, high, medium, etc.). Managers can exploit the simulation environment to approximate the distribution of customer opinions depending on some management choices. On the one hand, the manager could be interested in maintaining the initial distribution of opinions during implementation of specific management changes, for instance by putting in place a spending review. On the other hand, they might desire to improve initial distribution by applying specific business strategies, for instance: increasing the quality of services provided, and so on.

Figure 2 shows an overview of the simulation process aimed to evaluate choice impact in terms of Loyalty and/or Satisfaction. The process could be summed up as follows:

(1) *Identifying Opinion Change*, managers on the basis of expertise could identify which and to what extent, ideal customer opinions will change as consequence of specific management choices. For each antecedent variable (e.g., *Affective Commitment, Calculative Commitment*, etc. in the Fig. 2) of the underlying hypothesis, managers could use a slider to specify change of the customer opinions. For instance, incrementing fees of a service will probably impact the *Calculative Commitment* on the part of a customer. Consequently, managers will simulate a decreasing percentage of *Calculative Commitment* opinions. Figure 2 shows antecedent variables changed by using the sliders (i.e., Calculative Commitment). At the end of this step, retrieved opinions will be impacted and their values in terms of *Loyalty* or *Customer Satisfaction* estimated by classifying data by means of fuzzy IF-THEN rules extracted performing Fuzzy Data Analysis (see Sect. 4);

(2) *Evaluation of Impact*, taking into account the proposed change, the simulation engine identifies consequents (on which the proposed theoretical framework has been implemented) impacted by the change (e.g., *Loyalty* in the

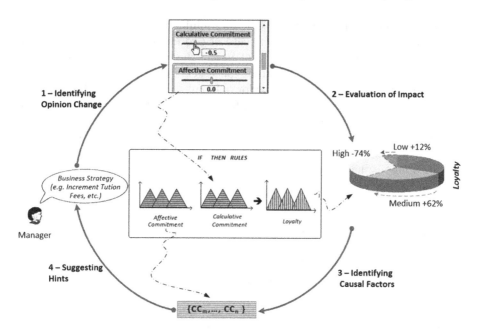

Fig. 2. Overview of the simulation process.

Fig. 2). Furthermore, the simulation engine exploits embedded IF-THEN fuzzy rules to re-classify collected customers' opinions in order to estimate how many customers have high level of loyalty (with respect to the selected consequent variables);

(3) *Identification of causal factors*, the simulation engine exploiting underlying hypothesis identifies concurring factors that influence customer distribution. These are the antecedents of extracted IF-THEN rules and could be used to mitigate the impact of population distribution change with respect to interesting variables (e.g., *Loyalty*);

(4) *Suggesting Hints*, the system returns to the manager which factors (identified before) could be changed to mitigate the impact of specific business strategy or management choice. For instance, relative to the decision of increasing tuition fees, the system could highlight that by improving the quality of services, the impact on the Loyalty will be smoothed.

6   Related Works

In order to achieve sustainable competition advantage in the market, it is necessary to provide and improve customer satisfaction and loyalty. In literature, many studies deal with the antecedents of satisfaction and loyalty [5].

Usually, Structural Equation Modeling (SEM) is widely applied to perform confirmatory analysis about relationships between customer satisfaction and loyalty (causal analysis or path analysis). Nevertheless, some studies compare or take advantage complementing SEM with data mining techniques (ANFIS, Fuzzy Decision Tree, and so on), such as the case of two emblematic works proposed in [50,51]. In [50] the authors argue that in the theoretical model of SEM, the key variables of interest are usually latent constructs (abstract psychological concepts), and so we can observe the behavior of latent variables only indirectly, and imperfectly, through their effects on manifest variables. Moreover, SEM requires a set of assumptions such as independent observations, random sampling of respondents, and the linearity of all relationships which are not achieved most frequently in practice [52]. The surveys used for such studies contain questions involving opinions, habits, attitudes of individuals. They often appear to be subjective and vague. The commonly used 5- or 7-point Likert scales can lead to vague answer which do not accurately portray respondent perception of the situation [53]. Such perceptions may have nonlinear relationships to each other, and so, they cannot be directly estimated in SEM [54]. Specifically, in [50] a comparison between SEM and ANFIS (i.e., adaptive neuro-fuzzy inference systems) has been performed. The authors argue that, unlike SEM, ANFIS can deal with nonlinear relationships in the forecasting of value innovation models using vague data in the questionnaire surveys. Unlike [50], the theoretical framework proposed in this work integrates the causal relationships resulting by the SEM to train fuzzy classifier which a simulation environment relies on to support contextual decision making. On the other hand, the problem investigated in [51] is developing post-analysis method for customer satisfaction and customer loyalty analysis, integrating data mining tools with the SEM technique to produce results useful for creating customer strategies. The resulting model is a kind of customer satisfaction evaluation that uses data mining advantages of classification algorithms (discovering unknown patterns) and cause-and-effect modeling advantage of SEM. Analogously with the work described in [51], our study proposes to exploit data mining techniques to define an integrated simulation environment in order to put the results of fuzzy data analysis into action to support simulation during the complex process of management decision making.

The theoretical framework proposed here hybridizes these paradigms of knowledge-based [55] and data-driven [56–58], DSS performing analysis of survey questionnaire to tailor decision support according to the context implied by the collected customer opinions. Recently, opinions are becoming a mainstream source of information in many of our day-to-day decision making tasks. In particular, the ability to efficiently utilize opinions (and data in general) to support all sorts of decision making tasks would be greatly useful in terms of adopting business strategy [59,60].

7   Conclusions and Future Works

The proposed theoretical framework defines a workflow to address the definition of context sensitive simulation environment to support decision making in the management field. Specifically, fuzzy data analysis of customer opinions enables the automatic tailoring of causal relationships, confirmed by performing SEM, in order to build a decision support system adherent to the context of application. In the area of management, this represents an innovative approach enabling *what-if* analysis to quantify and qualify correlations among statistical factors, i.e., extracting fuzzy rules from the raw data, customer opinions. Unlike other works that use only SEM to perform causal analysis, our study starts from the confirmed causal relationships to put in place in management decision making via a context sensitive simulation environment.

Future works will be aimed at substantiate the proposed theoretical framework in various public-founded universities in order to provide experimental results relative to performance using the fuzzy classifier and the effectiveness of the simulation environment. Furthermore, other outcomes of cross-fertilization of know-how in both management and computer science disciplines will hopefully, be addressed in the next future.

References

1. Chesbrough, H.: Open innovation: a new paradigm for understanding industrial innovation. In: Open Innovation: Researching A New Paradigm, pp. 1–12 (2006)
2. Assogna, P.: Innovation through cross-fertilization. In: NGEBIS Short Papers, pp. 12–17 (2013)
3. Björkdahl, J.: Technology cross-fertilization and the business model: the case of integrating icts in mechanical engineering products. Res. Policy **38**(9), 1468–1477 (2009)
4. Cusumano, M.A., Elenkov, D.: Linking international technology transfer with strategy and management: a literature commentary. Res. Policy **23**(2), 195–215 (1994)
5. Gustafsson, A., Johnson, M.D., Roos, I.: The effects of customer satisfaction, relationship commitment dimensions, and triggers on customer retention. J. Mark. **69**(4), 210–218 (2005)
6. Allen, N.J., Meyer, J.P.: The measurement and antecedents of affective, continuance and normative commitment to the organization. J. Occup. Psychol. **63**(1), 1–18 (1990)
7. Meyer, J.P., Allen, N.J.: Commitment in the Workplace: Theory, Research, and Application, Advanced Topics in Organizational Behavior. SAGE Publications (1997). http://books.google.it/books?id=jn4VFpFJ2qQC
8. De Witte, H., Näswall, K.: Objective'vssubjective'job insecurity: consequences of temporary work for job satisfaction and organizational commitment in four european countries. Econ. Ind. Democracy **24**(2), 149–188 (2003)
9. Morgan, R.M., Hunt, S.D.: The commitment-trust theory of relationship marketing. J. Mark. **58**, 20–38 (1994)
10. Werner, C.M., Turner, J., Shipman, K., Twitchell, F.S., Dickson, B.R., Bruschke, G.V., Wolfgang, B.: Commitment, behavior, and attitude change: an analysis of voluntary recycling. J. Environ. Psychol. **15**(3), 197–208 (1995)
11. Kwon, H.H., Trail, G.T., et al.: A reexamination of the construct and concurrent validity of the psychological commitment to team scale. Sport Mark. Q. **12**(2), 88–93 (2003)
12. Bansal, H.S., Irving, P.G., Taylor, S.F.: A three-component model of customer to service providers. J. Acad. Mark. Sci. **32**(3), 234–250 (2004)

13. Casper, J.M., Gray, D.P., Stellino, M.B.: A sport commitment model perspective on adult tennis players participation frequency and purchase intention. Sport Manag. Rev. **10**(3), 253–278 (2007)
14. Swimberghe, K., Sharma, D., Flurry, L.: An exploratory investigation of the consumer religious commitment and its influence on store loyalty and consumer complaint intentions. J. Consum. Mark. **26**(5), 340–347 (2009)
15. Bilgihan, A., Okumus, F., Cobanoglu, C., et al.: Generation y travelers commitment to online social network websites. Tour. Manag. **35**, 13–22 (2013)
16. Lyons, S.T., Duxbury, L.E., Higgins, C.A.: A comparison of the values and commitment of private sector, public sector, and parapublic sector employees. Public Adm. Rev. **66**(4), 605–618 (2006)
17. Gould-Williams, J.: The effects of high commitmenthrm practices on employee attitude: the views of public sector workers. Public Adm. **82**(1), 63–81 (2004)
18. Knapp, T., Fisher, B., Levesque-Bristol, C.: Service-learning's impact on college students' commitment to future civic engagement, self-efficacy, and social empowerment. J. Community Pract. **18**(2–3), 233–251 (2010)
19. Garbarino, E., Johnson, M.S.: The different roles of satisfaction, trust, and commitment in customer relationships. J. Mark. **63**, 70–87 (1999)
20. Hennig-Thurau, T., Gwinner, K.P., Gremler, D.D.: Understanding relationship marketing outcomes an integration of relational benefits and relationship quality. J. Serv. Res. **4**(3), 230–247 (2002)
21. Geyskens, I., Steenkamp, J.-B.E., Scheer, L.K., Kumar, N.: The effects of trust and interdependence on relationship commitment: a trans-atlantic study. Int. J. Res. Mark. **13**(4), 303–317 (1996)
22. Fecikova, I.: An index method for measurement of customer satisfaction. TQM Mag. **16**(1), 57–66 (2004)
23. Oliver, R.: Satisfaction: A Behavioral Perspective on the Consumer. McGraw-Hill series in marketing. McGraw Hill (1997). http://books.google.it/books?id=rjGW QgAACAAJ
24. Oliver, R.L., Swan, J.E.: Consumer perceptions of interpersonal equity and satisfaction in transactions: a field survey approach. J. Mark. **53**, 21–35 (1989). http://www.jstor.org/stable/1251411
25. Elliott, K.M., Shin, D.: Student satisfaction: an alternative approach to assessing this important concept. J. High. Educ. Policy Manag. **24**(2), 197–209 (2002). doi:10.1080/1360080022000013518. http://dx.doi.org/10.1080/1360080022 000013518
26. DeShields, O.W., Kara, A., Kaynak, E.: Determinants of business student satisfaction and retention in higher education: applying herzberg's two-factor theory. Int. J. Educ. Manag. **19**(2), 128–139 (2005). doi:10.1108/09513540510582426. http://www.emeraldinsight.com/doi/pdf/10.1108/09513540510582426
27. Marzo Navarro, M., Pedraja Iglesias, M., Rivera Torres, P.: A new management element for universities: satisfaction with the offered courses. Int. J. Educ. Manag. **19**(6), 505–526 (2005). doi:10.1108/09513540510617454. http://www.emeraldinsight.com/doi/abs/10.1108/09513540510617454
28. Zins, A.H.: Relative attitudes and commitment in customer loyalty models. Int. J. Serv. Ind. Manag. **12**(3), 269–294 (2001). doi:10.1108/EUM0000000005521. http://www.emeraldinsight.com/doi/pdf/10.1108/EUM0000000005521
29. Gremler, D.D., Brown, S.W., et al.: Service loyalty: its nature, importance, and implications. In: Advancing Service Quality: A Global Perspective, pp. 171–180 (1996)

30. Pritchard, M.P., Howard, D.R.: The loyal traveler: examining a typology of service patronage. J. Travel Res. **35**(4), 2–10 (1997). doi:10.1177/004728759703500417. http://jtr.sagepub.com/content/35/4/2.full.pdf+html
31. Fullerton, G.: The impact of brand commitment on loyalty to retail service brands. Can. J. Adm. Sci./Rev. Can. des Sci. de l'Administration **22**(2), 97–110 (2005)
32. Mattila, A.S.: How affective commitment boosts guest loyalty (and promotes frequent-guest programs). Cornell Hotel Restaurant Adm. Q. **47**(2), 174–181 (2006)
33. Evanschitzky, H., Iyer, G.R., Plassmann, H., Niessing, J., Meffert, H.: The relative strength of affective commitment in securing loyalty in service relationships. J. Bus. Res. **59**(12), 1207–1213 (2006)
34. Hennig-Thurau, T.: Customer orientation of service employees: Its impact on customer satisfaction, commitment, and retention. Int. J. Serv. Ind. Manag. **15**(5), 460–478 (2004)
35. Dimitriades, Z.S.: Customer satisfaction, loyalty and commitment in service organizations: some evidence from greece. Manag. Res. News **29**(12), 782–800 (2006)
36. Andreassen, T.W., Lindestad, B.: Customer loyalty and complex services: the impact of corporate image on quality, customer satisfaction and loyalty for customers with varying degrees of service expertise. Int. J. Serv. Ind. Manag. **9**(1), 7–23 (1998)
37. Bolton, R.N.: A dynamic model of the duration of the customer's relationship with a continuous service provider: the role of satisfaction. Mark. Sci. **17**(1), 45–65 (1998)
38. Patterson, P.G., Spreng, R.A.: Modelling the relationship between perceived value, satisfaction and repurchase intentions in a business-to-business, services context: an empirical examination. Int. J. Serv. Ind. Manag. **8**(5), 414–434 (1997)
39. Monecke, A., Leisch, F.: sempls: structural equation modeling using partial least squares
40. Bollen, K.A.: Structural Equations with Latent Variables. Wiley, New York (2014)
41. Yoon, Y., Gursoy, D., Chen, J.S.: Validating a tourism development theory with structural equation modeling. Tour. Manag. **22**(4), 363–372 (2001)
42. Gefen, D., Straub, D., Boudreau, M.-C.: Structural equation modeling and regression: guidelines for research practice. Commun. Assoc. Inf. Syst. **4**(1), 7 (2000)
43. Fornell, C., Larcker, D.F.: Structural equation models with unobservable variables and measurement error: algebra and statistics. J. Mark. Res. **18**, 382–388 (1981)
44. Jairak, R., Praneetpolgrang, P.: Using fuzzy cognitive map based on structural equation modeling for designing optimal control solution for retaining online customers
45. Bezdek, J.C.: Pattern Recognition with Fuzzy Objective Function Algorithms. Kluwer Academic Publishers, Norwell (1981)
46. Yu, H.-C., Lee, Z.-Y., Chang, S.-C.: Using a fuzzy multi-criteria decision making approach to evaluate alternative licensing mechanisms. Inf. Manag. **42**(4), 517–531 (2005)
47. Xie, X.L., Beni, G.: A validity measure for fuzzy clustering. IEEE Trans. Pattern Anal. Mach. Intell. **13**(8), 841–847 (1991). doi:10.1109/34.85677
48. Hoppner, F., Klawonn, F., Kruse, R., Runkler, T.: Fuzzy cluster analysis: methods for classification, data analysis and image recognition. J. Oper. Res. Soc. **51**(6), 769 (2000)
49. Mamdani, E.H.: Application of fuzzy logic to approximate reasoning using linguistic synthesis. IEEE Trans. Comput. **C−26**(12), 1182–1191 (1977). doi:10.1109/TC.1977.1674779

50. Ho, Y.-C., Tsai, C.-T.: Comparing ANFIS and SEM in linear and non-linear forecasting of new product development performance. Expert Syst. Appl. **38**(6), 6498–6507 (2011). doi:http://dx.doi.org/10.1016/j.eswa.2010.11.095. http://www.sciencedirect.com/science/article/pii/S0957417410013333

51. Aktepe, A., Ersz, S., Toklu, B.: Customer satisfaction and loyalty analysis with classification algorithms and structural equation modeling. Comput. Ind. Eng. (2014). doi:http://dx.doi.org/10.1016/j.cie.2014.09.031. http://www.sciencedirect.com/science/article/pii/S0360835214002988

52. Şen, Z., Altunkaynak, A.: Fuzzy system modelling of drinking water consumption prediction. Expert Syst. Appl. **36**(9), 11745–11752 (2009). http://www.scopus.com/inward/record.url?eid=2-s2.0-67349274436&partnerID=40&md5=34188682fe3c440ae84d2fefe0fad1d0

53. Deng, W.-J., Pei, W.: The impact comparison of likert scale and fuzzy linguistic scale on service quality assessment. Chung Hua J. Manag. **8**(4), 19–37 (2007)

54. Hair, J.: Multivariate Data Analysis. Prentice Hall (1998). http://books.google.it/books?id=-ZGsQgAACAAJ

55. Suresh, S., Naidu, M., Kiran, S.A.: An xml based knowledge-driven decision support system for design pattern selection. Int. J. Res. Eng. Technol. (IJRET) 1 (3)

56. Grond, M., Morren, J., Slootweg, H.: Requirements for advanced decision support tools in future distribution network planning. In: 22nd International Conference and Exhibition on Electricity Distribution (CIRED 2013), pp. 1–4 (2013). doi:10.1049/cp.2013.1050

57. Fenza, G., Furno, D., Loia, V., Senatore, S.: Approximate Processing in Medical Diagnosis by Means of Deductive Agents, Chap. 23, pp. 633–657. doi:10.1142/9789814329484_0023. http://www.worldscientific.com/doi/abs/10.1142/9789814329484_0023

58. Maio, C.D., Fenza, G., Gaeta, M., Loia, V., Orciuoli, F.: A knowledge-based framework for emergency DSS. Knowl. Based Syst. **24**(8), 1372–1379 (2011). doi:http://dx.doi.org/10.1016/j.knosys.2011.06.011. http://www.sciencedirect.com/science/article/pii/S0950705111001262

59. Entani, T.: Process of group decision making from group and individual viewpoints. In: 2012 Joint 6th International Conference on Soft Computing and Intelligent Systems (SCIS) and 13th International Symposium on Advanced Intelligent Systems (ISIS), pp. 1322–1326 (2012). doi:10.1109/SCIS-ISIS.2012.6505055

60. Alonso, S., Herrera-Viedma, E., Chiclana, F., Herrera, F.: Individual and social strategies to deal with ignorance situations in multi-person decision making. Int. J. Inf. Technol. Decis. Making **08**(02), 313–333 (2009). doi:10.1142/S0219622009003417. http://www.worldscientific.com/doi/pdf/10.1142/S0219622009003417

Improved Recommendation System Using Friend Relationship in SNS

Qing Liao[1,3(✉)], Bin Wang[1,3], Yanxiang Ling[1,3], Jingling Zhao[2], and Xinyue Qiu[2]

[1] School of Information and Communication Engineering,
Beijing University of Posts and Telecommunications, Beijing 100876, China
[2] School of Computer Science, Beijing University of Posts and Telecommunications,
Beijing 100876, China
jing_lingzh@sina.com, 365562443@qq.com
[3] Beijing Key Laboratory of Network System Architecture and Convergence,
Beijing 100876, China
qingliao@hotmail.com, wangbinjs@126.com, cdxustb@163.com

Abstract. With the rapid development of the Internet, SNS services and 3G commercial mobile applications there have been tremendous opportunities although the development of SNS is very short in China, and the social web game is in the early stage of development. Because of massive users, the potential commercial value of Chinese SNS is still a great mining space. However, a relatively large defects is the precipitation and accumulation on content. The dynamic of friends will affect our own decisions largely, it is favorable for the activity of SNS to increase the number of friends. We have improved the existing models, and conduct experiments to validate it and compare it with previous methods.

Keywords: SNS · Context · Matrix decomposition model · Recommendation

1 Introduction

Since the birth of the search engines Google, we are discussing what is the next gold mine of the Internet. Now, almost everyone agrees that it is the social network. According to the report, Internet users almost take 22 % of the time in the social media and social networks. Facebook and Twitter as a representative of two different types of social networks, are pioneers in the Internet sector SNS. Social networks have two most important factors: social relationships and social data information, social relationships also known as friendships. As we all known, dynamics of our friends on the SNS will largely affect our own decisions. For example, we show their new shoes in the SNS, it is likely to affect our friends. This kind of influence is greater than the power of the seller. That is to say, friends can spread the "trust". Based on this, the SNS pay more attention to the topology relationship of the user's friends.

According to the findings of the U.S. investigation agency Nielsen, it shows that more than 90 % of users believe in what their friends have recommended, and then 70 % of users believe the score of the product that other users rated on a commodity in the Internet. Therefore, what friends recommend is important to increase user's trust recommendation result. Nielsen once made a personalized ads experiments with Facebook,

© Springer-Verlag Berlin Heidelberg 2015
N.T. Nguyen et al. (Eds.): Transactions on CCI XIX, LNCS 9380, pp. 17–31, 2015.
DOI: 10.1007/978-3-662-49017-4_2

and do the ABtest. Nelson display the same brand in three different ways of advertising, the first is to tell the users of Facebook that there are 50000 users pay attention to the brand, the second is to tell the users of Facebook that how many friends pay attention to this brand, and the third is to tell users what friends pay attention to the brand. According to the analysis of AD click results, the effect of the third is better than that of the second, the second is better than the first. So increasing the number of friends is conducive to increase the activity of the social product, which increases the effect of advertising.

2   Recommended Types of Social Networks

2.1   The Representation and the Characteristics of Social Network Data

Social networks (SNS) define the relationship between users, according to the definition of figure, we represent SNS by G(V,E,W), and V represents the vertices, E represents the set of edges. If the user Va and Vb have social relationships (which is often said friends relationship), then there is an edge e(Va,Vb) connecting these two users. W(Va,Vb) represent the weight of edge which shows the degree of relationship between friends, and most can be set to 1. Today there are two kinds of SNS, one is Facebook that represents the bidirectional relationship of friends, e(Va,Vb) which is an undirected edges can describes it. Another is twitter that represents the unidirectional relationship between friends, so directed edges represent the relationships.

Out(u) is a set of vertices which vertex u point to in the Figure G, in(u) is a set of vertices which point to vertex u. Generally, there are three different social network data:

(1) SNS of bidirectional relationship of friends; for example, Facebook and renren, $out(u) = in(u)$.
(2) SNS of unidirectional relationship of friends; the typical representative is Twitter and Sina weibo.
(3) SNS of community groups, the relationship between SNS users is uncertain, however, this data includes the user data belonging to different communities. For example, douban group, which belonging to the same group may represent the users have similar interest.

There are two types of recommendations based on social networks, one is recommended items, items can be advertising, news and more. Another one is friend's recommendation.

2.2   Recommended Items Based on Social Networks

Many websites recommend products to users by SNS, for example, Amazon recommends product to users based on what their friends like.

Recommendation from friends can increase the recommendation trust. Friends of the users on the SNS tend to be trust, at the same time the users are disgusted with the results of computer calculated. For example, Web site recommend "naruto" to users, if it is based on the item with the filtering algorithm, the recommended reason may be that "naruto" and "one piece" is similar. But if based on SNS recommended, the recommended reason is that there are 10 friends of the user like "naruto", apparently users tend to accept the second reason.

Ease the problem of cold starting. When a user login weibo on e-commerce sites, we can get the user's list of friends from weibo, then get the consumption records of their friends' and find out the right item to recommend to the current user.

Of course, the method of item recommendation based on SNS also can produce bad case. Because the user's friends relationship are not usually based on common interests (relatives, classmates friends are also friends). So the different interest of user's friends, make recommendation accuracy and recall rate of the algorithm decreased.

The behavior of users that buy goods can be shown by users diagram and Items diagram, and the combination of the figure and users of social network diagrams together is a bipartite graph. Figure 1 is an example of a bipartite graph combines social networking and user objects. If a user u produce a behavior for goods i, then there is the edge between two nodes. For example, a user buy objects a and objects e, If user B and D are friends, then there will be an edge connecting the two users, the user A in Fig. 1 and the user B, D are friends.

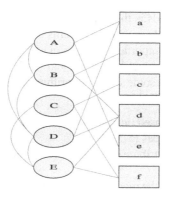

Fig. 1. The combine of social networking and user objects graph

The similarity between users also means the weight between users, which can be defined as α. The weight between the user and the items may be defined as the degree of the product user liked, which can be defined as β. The value of α and β are based on the training data. After determining the edge weight, we can use PersonalRank figure sorting algorithms to generate recommendations for each user.

In a social network, in addition to the common and directly relationship between the user's social network, there is another relationship which the user belong to the same group (such as douban interest group). At this point, the map mode will be more simple. We can modeling the relationship between the group and the user's friends. Figure 2, adding a node that represents a community, and if the user belongs to a community, there is an edge between the group nodes and the user nodes.

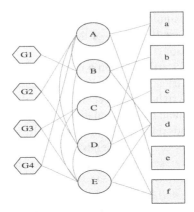

Fig. 2. Integration of two social network graph model information

2.3 Recommend of SNS Friends

The relationship of good friends is the important part in the SNS, if the user's friends are very few, it will not be able to realize the benefits of socialization. Therefore, recommending friends is one of the important application of SNS. The recommended system of friends is to recommend product by friends and the interact daily record of user's behavior, first briefly introduce some simple algorithm, and put forward our own recommendation algorithm of SNS friends.

(1) the matching algorithm based on content
     User content has many types [5].
     a. User Demographics attribute content, which includes age, sex, work, residence, education and so on.
     b. User interest which is the most important content, including user favorite items and published remarks.
     c. User context, including GPS location, mood, weather, time, etc.
     Based on the above information, we calculate the similarity between users, and then recommend friends. In fact, these three pieces of information called user context information. In the proposed algorithm, the full use of context information can improve the accuracy of the recommended.
(2) recommendations based on common interest of friends
     In SNS friends, the users become a friend based on the common interest, they do not care if they knew each other in the real world. Therefore, recommendations based on common interests of friends is needed in sina microblogging.
     User-based collaborative filtering algorithms can be used to calculate the user similarity, the main similarity analysis is based on the same score on the same items, for example, when users comment or forwarding the same microblogging, then we believe that user likes this microblogging.

(3) Recommendations of fiends based on a social network graph.

We analysis social networking by the graph model, especially new users, we recommend friend by user groups, or recommend friends of friends by taking advantage of the spread of the map, here is a recommendation algorithm of friends based on social networking.

We can calculate their similarity degree based on the proportion of common friends for user U and V:

$$w_{out}(u, v) = \frac{|out(u) \cap out(v)|}{\sqrt{|out(u)| \, |out(v)|}}$$

(1)

out(u) means a set of friends collections that user U point to, in(u) means a set of friends collections who point to user U, approximation defined as in(u) can be represent as follow:

$$w_{in}(u, v) = \frac{|in(u) \cap in(v)|}{\sqrt{|in(u)| \, |in(v)|}}.$$

(2)

3 Improved Recommendation of SNS Friends: Matrix Decomposition Model on Context

As is known to all, SNS is a gold mine, especially the rapid development in recent years, as the representative of SNS, sina Weibo and tencent weibo daily output plenty of information in China. So in order to get more accurate information and people, to build a good friend recommendation system is indispensable. Without the users' activity, SNS service is a "dead city". Traditional recommendation systems, such as user-based CF and Item-based CF, both only taking the user or objects into account, but ignoring the property of users or items. As a result, it will be difficult to grasp the true preferences of the user, and the recommendation accuracy can not be guaranteed. Therefore, we must fully consider the context to improve accuracy of recommendation. This paper will present an improved algorithm and model, which can improve the accuracy of recommendation with the integration of the context of the user's, the context of time and the

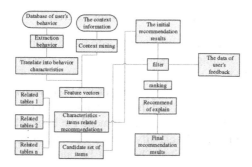

Fig. 3. Recommend recommendation system model based on context

matrix decomposition model. Figure 3 is our recommendation system model based on context.

Simply explain the role of the various modules

a. The Database of user behavior. Users can create many different kinds of behaviors, such as user browse items, click on the link address of items, add items to shopping cart, purchase items, to score, etc. These behaviors can make guidance to the recommend of users. Each kind of behavior can be used as a dimension of user behavior feature vector, through training and learning, and get the final weight of behavior.

b. The context information. A good recommendation system, must consider the context, this context can be the user's population information, attributes of the item, the time context, site context, etc. But in some cases, certain context is not obvious, for example, we only know the IP address of the user, in order to get the user's location, you need the mining module to analysis the context.

c. Related tables. Related tables can be understood as a configuration file, stored the weight on a characteristic value.

d. Characteristics - items related recommendations. After the feature vector of the customers, according to the configuration of related tables, we get the initial list of recommended items offline.

e. Filter module. To filter out the following items:
    (1) The items which the user has behaved.
    (2) The items besides candidates. Such as user search clothes in taobao, rules must be men's clothes, you need to get rid of the clothes which are not men's.
    (3) Poor quality items. To get a lot of bad review, you need to get rid of these products.

f. Ranking module. Ranking module is to ensure the novelty of the item, which would require to drop the weight of popular items. In addition, need to ensure diversity, and this need to cover the user's various interest. Finally, to guarantee time diversity, meaning that the user does not want to see the same recommendation every day, so the recommendation system should focus on recent actions of users.

g. User feedback. User feedback can help recommend system to get the user's interest. We can predict that whether the user can click the recommendation results based on the click models.

3.1 The Introduction of Matrix Decomposition Model

The essentially of Matrix decomposition model is to find the potential characteristics of users, in the field of text mining, also known as latent factor model (LFM) [1–4]. Since the Netflix Prize, the LFM has gradually become the powerful tool of the recommendation system. For the User - based CF, you first need to find users who read the same books with the recommended users, then recommend other books which that users liked. For the Item-based CF, we need to recommend the book which is similar to that books they have read. First of all, classify items by characteristics, for users, get his interest categories, then we select items he might be interested. Classification can be classified by the editorial staff, but it is difficult for us to determine the right of an item.

And because matrix decomposition model take use of automatic clustering based on the statistics of user's behavior, it becomes a better solution to the problem of classification weights.

(1) Singular value decomposition

Recommendation system can extract a set of potential (hidden) factors from the rating model by using matrix factorization, and describe users and items by these factors vectors. In the late 1980 s, using the ideas of potential semantic lots factor has been successfully applied in the field of information retrieval. Deerwester proposed using singular value decomposition (SVD) technique to find potential factors in the document. In information retrieval, this kind of latent semantic analysis (LSA) technology is also classified as latent semantic retrieval (LSI).

Problems in the field of information retrieval are usually based on the user's query words to find a set of documents. Document and the user query is resolved as a vector of words. This retrieval method can't solve the synonyms in the document or the query words. SVD regards that highly relevant and appear together words as individual factor, and make the usually large document vector matrix apart into smaller order approximation matrix. So we can retrieve relevant documents in the case that do not contain user query words based on LSI.

In the field of recommender systems, people pay more attention to the latent semantic model and matrix decomposition model. In fact, the nature is dimension reduction techniques, and score matrix completion.

User's scores on items can be expressed as a rating score matrix R, and R [u] [i] indicates that the user's scores on item i. However, the user can not comment on all items, so many elements of this matrix is empty, such as if a user does not score on an item, then they would have to predict whether the user is able to score for this item and assess how many points.

For how to complete a matrix, there are a lot of research in the history. There are many ways to complete an empty matrix. We want to find a completion of matrix perturbation method which have a minimum disturbance. So which is the minimum disturbance for matrix? It is generally believed that if the eigenvalues of the matrix after completion and before the completion of eigenvalues were similar, we said that the disturbance for matrix is minimum.

The first matrix decomposition model is SVD [6], there are m users and n items, and scoring matrix $R^{m \times n}$. First we need to complement the missing items' score in the matrix, by the way of using the global average or the average of users and items, then get the completion matrix R'. Then we decompose R' into the following form:

$$R' = U^T SV \tag{3}$$

$U \in M^{k \times m}$ and $V \in M^{k \times n}$ are two orthogonal matrix, $S \in M^{k \times k}$ is a diagonal matrix, each of the diagonal elements are the singular values. In order to reduce the dimension of the R', we take the largest singular values and composed the diagonal matrix S_f, and get a scoring matrix with dimensionality reduction:

$$R'_f = U_f^T S_f V \tag{4}$$

SVD is used by the early recommendation system, but this method has the disadvantage:

(1) This method first need to complement the sparse scoring matrix with a simple method. So matrix will become a dense matrix storage, and take the great storage space.

(2) The calculation of this method is large, especially in a dense matrix. The actual recommendation system is on the hundreds of millions of users and even one hundred million level of goods, apparently this method can not be used.

We'll decompose the score matrix R into a low latitudes matrix multiplication:

$$\hat{R} = P^T Q \tag{5}$$

$P \in M^{f \times m}$ and $Q \in M^{f \times n}$ are two dimension reduction matrix. P represents user implicit characteristic vector, Q represents the hidden feature vector of items, f represents the number of users' or items' hidden features, we can define our own, in theory, the greater the accuracy is higher, but the memory consumption space is greater. $\hat{R}(u, i)$ means the score of the item i from user u.

$$\hat{R}(u, i) = \hat{r}_{ui} = \sum_f p_u q \tag{6}$$

Loss function is as followed:

$$
\begin{aligned}
C(p, q) &= \sum_{(u,i) \in train} \left(\hat{r}_{ui} - r_{ui} \right)^2 \\
&= \sum_{(u,i) \in train} \left(r_{ui} - \sum_{f=1}^{F} p_{uf} q_{if} \right)^2
\end{aligned}
\tag{7}
$$

Then we add prevent fitting items $\lambda \left(\|p_u\|^2 + \|q_i\|^2 \right)$, so the formula is as followed:

$$
\begin{aligned}
C(p, q) &= \sum_{(u,i) \in train} \left(\hat{r}_{ui} - r_{ui} \right)^2 \\
&= \sum_{(u,i) \in train} \left(r_{ui} - \sum_{f=1}^{F} p_{uf} q_{if} \right)^2 + \lambda \left(\|p_u\|^2 + \|q_i\|^2 \right)
\end{aligned}
\tag{8}
$$

Then we optimize parameters with stochastic gradient descent algorithm, SGD through continuous iteration strategy, reduce the iteration error, stopped until the compressed to a certain range of allowable error.

$$\frac{\partial C}{\partial q_{if}} = -2p_{uk} + 2\lambda q_{ik}$$

$$\frac{\partial C}{\partial p_{uf}} = -2q_{ik} + 2\lambda p_{uk} \tag{9}$$

The recursive formula is as followed:

$$p_{uf} = p_{uf} + \alpha \left(q_{ik} - \lambda p_{uk} \right)$$

$$q_{if} = q_{if} + \alpha \left(p_{uk} - \lambda q_{ik} \right) \tag{10}$$

α is the learning rate, it needs to get through trial. A scoring system has some inherent attributes which have nothing to do with users and items, and some properties of the items also has nothing to do with the users. So, someone put forward the following matrix decomposition model, the prediction formula is:

$$\hat{r}_{ui} = \mu + b_u + b_i + p_u^T q_i \tag{11}$$

μ: The global average of all the records of score of the training sets. In different sites, because the site location and items are different, the distribution of the overall site's grade will also show some differences. Global average can represent the impact on user ratings with the website itself.

b_u: User bias. It reprents the unrelated factors with items in the user's rating habits. For example, some users are demanding, some users are more tolerante, then it will appear different ratings for the same items.

b_i: Item bias. It represents the irrelevant factors with users in the ratings of items. For example, some items itself has high quality, so the score is relatively high, and poor quality of some items, relative score will be low.

Modified ItemCF prediction algorithm is as follows:

$$\hat{r}_{ui} = \frac{1}{\sqrt{|N(u)|}} \sum_{j \in N(u)} w_{ij} \tag{12}$$

w_{ij} can be optimized by the following optimizing loss function:

$$C(w) = \sum_{(u,i) \in T} \min \left(r_{ui} - \sum_{j \in N(u)} w_{ij} r_{uj} \right)^2 + \lambda w_{ij}^2 \tag{13}$$

However, w matrix of the model is very dense, for storage, it will cost a lot of memory space. In addition, if there are n items, then the parameters of the model number is n^2, the number of parameters is large, so it is easy to cause the fitting. Koren[8] proposed that we should decompose the w matrix, the parameter number decreased to $2 * n * F$, model is as follows:

$$\hat{r}_{ui} = \frac{1}{\sqrt{|N(u)|}} \sum_{j \in N(u)} x_i^T y_j$$

$$= \frac{1}{\sqrt{|N(u)|}} x_i^T \sum_{j \in N(u)} y_j \tag{14}$$

We will add the LFM and the above model, so as to get the following models:

$$\hat{r}_{ui} = \mu + b_u + b_i + p_u^T q_i$$

$$+ \frac{1}{\sqrt{|N(u)|}} x_i^T \sum_{j \in N(u)} y_j \tag{15}$$

Koren in order not to increase too much parameter to fitting, we define $x = q$

$$\hat{r}_{ui} = \mu + b_u + b_i$$

$$+ q_i^T \left(p_u + \frac{1}{\sqrt{|N(u)|}} \sum_{j \in N(u)} y_j \right). \tag{16}$$

3.2 Improved Model

Compared with traditional recommendation problem, recommend friends for weibo is more challenging. Weibo users have rich interaction records. Such as tweeting, forwarding, commentary, attention/concern, @, each kind of interaction can be regarded as some preferences of users. Moreover, the recommended item is a special kind of users, and have the same behavior with ordinary users. The most important, the environment context and time context of users also determine the quality of friends recommend, user context can be the attributes of users, such as age, occupation, company, etc. while the time context will take into account the currently preferences of the users, because the preferences will change with time, so the time context is essential.

U: User sets, u means one of the users,|U|: the number of users.

I: Items set, |I| means one of the items, I: the number of the recommend item.

r_{ui}: It represents the score of items from users, for weibo friend recommendation, only two points, 1 means accepted and 0 means not accepted.

$\hat{r}_{ui}$: It represents predict score of items from users.

$R_{m \times n}$: It represents rating matrix of n items from m users.

Firstly the user's interest will change with time, for example, he likes football today, however he maybe interested in basketball in the next days. Secondly the user's behavior on SNS will be different during the day and night, for example, a user is working during the day, so because of time, he is likely to browse weibo quickly, but seldom pay attention to the recommended friends, on the contrary, the evening will have a lot of time to deal with weibo for details. Thirdly the popularity of the article may be reduced with time. Based on the above, this article will integrate the context in the LFM model.

We add the user's context into LFM model:

$$\hat{r}_{ui} = \mu + b_{ui}(t)\,\tilde{q}_i^T\,\tilde{p}_u \tag{17}$$

μ represents the global average score of all the records in the training sets, in different sites, because the different site location and different items, the overall site grade distribution will also show some differences. Global average can make impact on user ratings of the website itself. $b_{ui}(t)$ represents the offset item of users and items. $b_{ui}(t) = b_u(t) + b_i(t)$, $b_u(t)$ and $b_i(t)$ are the bias of the context, $\tilde{q}_i$ and $\tilde{p}_u$ represents the model of items and the user.

(1) The attributes of users and items integrate into the context of the article.
  We can dig out some users' interests and tastes with the context of the attributes of users and objects, such as age, sex, job, living area and so on. Users with similar age generally have similar hobbies, so we added two pairs of bias items: $b_{u,gen(i)}$, $b_{u,age(i)}$, $b_{gen(u),i}$, $b_{age(u),i}$.
  $b_{u,gen(i)}$, $b_{u,age(i)}$ indicates the preferences bias of recommended friends who has similar age and same sex, $b_{gen(u),i}$, $b_{age(u),i}$ represents preferences bias on users by some people who have a similar age or sex. Assuming the user's registered birth year is y, then do the following age categories:

$$age\,(u) = \begin{cases} 0 & y \le 1949 \\ ceil\left(\frac{y-1950}{3}+1\right) & 1950 \le y \le 2004 \\ 16 & y \ge 2004 \\ 17 & y \ \ is \ \ illegal \end{cases} \tag{18}$$

Sex categories:

$$gen(u) = \begin{cases} 0 & if \ \ sex \ \ is \ \ male \\ 1 & if \ \ sex \ \ is \ \ female \\ 2 & if \ \ sex \ \ is \ \ null \end{cases} \tag{19}$$

(2) The user activity bias into the context of the article.
  User activity means the number of microblogging that user send or forward, the higher the number, the more its activity. The bias degree of user's active is $b_{twnum(u)}$ twnum(u) = ceil(numTweet(u))/5 + 1 means the number that user send.

(3) The context of user tags.
  Tags marked users and the characteristics of recommendation, for example, a user labeled themselves as "IT", "Football", then we can make sure that this user is very interested in several of these field, so we can know the similarity between users with the similarity of tags. taglist(u) is the tag list of user u and taglist(v) is the tag list of objects that to be recommended. Then the similarity between them is expressed as:

$$sim_{tag}(u, v) = \frac{taglist(u) \cap taglist(v)}{taglist(u) \cup taglist(v)} \qquad (20)$$

The tag's bias of user u and user v is expressed as follow:

$$b_{tag} = \alpha_{u,v} sim_{tag}(u, v) \qquad (21)$$

Then we define the variable:

$$q_{u,tag} = \frac{\sum\limits_{n \in T(u)} Vec(n)}{|T(u)|} \qquad (22)$$

$$q_{i,tag} = \frac{\sum\limits_{n \in T(i)} Vec(n)}{|T(u)|} \qquad (23)$$

T(u) means the tag collection of user u, Vec(n) means n vectors of the tag.

(4) The content of microblogging and keywords.

When a user forwards or write weibo, essentially reflects the potential interest of the user. So we extract microblogging keywords, and use the similarity of keyword to analyze the association of users and items. Keywords can be extracted from the microblogging, forwarding microblogging, comments or the user's self-description. There is a weight for keywords k form each user means w(u,k).

Similarly, If a user u and item i have the same keywords, the keywords can be seen the common interest of both, we can get a formula which is similar to the bias of tag items. The tag list of user u is kwlist(u) and the tag list of objects that to be recommended is kwlist(v), then the similarity can be expressed: $sim_{kw}(u, v)$

We define the offset of keywords u and keywords v:

$$b_{kw} = \beta_{uv} sim_{kw}(u, v) \qquad (24)$$

β_{uv} is the weight of $sim_{kw}(u, v)$.

(5) The time integrate into the context.

The popularity of the recommended items will change over time. For example, a movie, it is possible becomes popular with the factors that from outside. So for the LFM model, items bias b_i can no longer be assumed to be constant term, while it should be a time-related functions. In addition, users bias b_u should be a time-related functions. We defines the user - Item Bias items:

$$b_{ui}(t) = b_i(t) + b_u(t) \qquad (25)$$

$b_{ui}(t)$ is the bias of items i from user u at time t, $b_u(t)$, $b_i(t)$ is the time offset function of users and items.

We discrete the time, each day is divided into time periods, each time period is one hour, hour(t) represents the number of hours a day (from 0 to 23). day(t) is the date in days (from 0, the maximum value of day(t) is 29). So the bias function of items is:

$$b_i(t) = b_i + b_{i,day(t)} + b_{i,hour(t)} \tag{26}$$

Based on the feedback time on recommendation from user, we introduce the time decay function which represent the degree of influence on scores:

$$c_{u,t} = \cfrac{1}{1 + \alpha_u \left(\cfrac{t - T_b}{T_e - T_b} \right)^2} \tag{27}$$

$c_{u,t}$ is the degree of influence on scores from users at time t, T_b is the start timestamp of training data, t is the current timestamp, T_e is the end timestamp of training data, α_u is the attenuation coefficient.

$$b_u(t) = b_u + \cfrac{1}{1 + \alpha_u (\cfrac{t - T_b}{T_e - T_b})^2}. \tag{28}$$

4 Experiment and Analysis

4.1 A Set of Training Data

The date is from tencent microblogging, the data is sampled from four hundred million tencent microblogging users about fifty days, which contains more than two hundred million active users, six thousand recommended users or information, more than three hundred million records and listening anction, more than seventy million training records, and more than thirty million test records.

4.2 Evaluation Standards

We suppose to recommend an ordered list of m items to a certain user, the user may select one of 0, 1 or more of the recommended items, then we define the average accuracy from information retrieval. If the denominator is zero, the result is set to 0. P(k) represents the accuracy at the cutoff point k on the list of items. When the k-th item is not selected, P(k) is zero. Here are a few examples:

(1) If the user selects the # 1, # 3, # 4 from the five recommended projects, its accuracy is

$$ap@3 = (1/1 + 2/3)/3 \approx 0.56$$

(2) If the user selects the # 1, # 2, # 4 from the four recommended projects, its accuracy is

$$ap@3 = (1/1 + 2/2)/3 \approx 0.67$$

(3) If the user selects the # 1, # 3 from the three recommended projects, its accuracy is

$$ap@3 = (1/1 + 2/3)/2 \approx 0.83$$

For the N users in the position n, the average accuracy is the accuracy of the mean average of each user, the following equation:

$$AP@n = \sum_{i=1,2,...N} ap_i@n_i/N.$$
(29)

4.3   The Results of Experiment

Table 1 shows the results of the recommendation system based on the context in the SNS:

Table 1. The results of the recommendation system based on the context in the SNS

type	describe	MAP@3
1	A	0.3443
2	1+ age	0.3606
3	2+ item	0.3611
4	3+ microblog	0.3621
5	4+ time	0.3697

A means the Basic matrix decomposition model.
Figure 4 shows the trends of change in the recommended results:

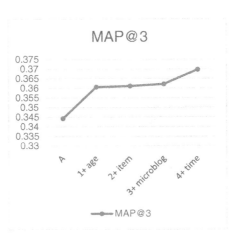

Fig. 4. The trends of change in the recommended results

As can be seen from Fig. 4, add the bias of time context into the system, it can effectively improve the recommendation accuracy.

5 Summary

Social networking can be the current gold mine. When users has more and more friends in the social network, SNS will be more active, which will bring more profits for social networking. Therefore recommending friends for users of SNS is also one of the important functions of the SNS. This artical apply the time context to the bias model of users and items, and excavate the context of user's activity. We extended matrix decomposition model with the demographic attributes of users and items, and verify the model with the data of tencent microblogging, at last achieved a good results on recommendation.

References

1. Koren, Y.: Factorization meets the neighborhood: a multifaceted collaborative filtering model. In: Proceedings of the 14th ACM SIGKDD International Conference on Knowledge Discovery and Data Mining, ACM, pp. 426–434 (2008)
2. Koren, Y.: Factor in the neighbors: scalable and accurate collaborative filtering. ACM Trans. Knowl. Discov. Data (TKDD) **4**(1), 1 (2010)
3. Koren, Y., Bell, R., Volinsky, C.: Matrix factorization techniques for recommender systems. Computer **42**(8), 30–37 (2009)
4. Hu, Y., Koren, Y., Volinsky, C.: Collaborative filtering for implicit feedback datasets. In: Eighth IEEE International Conference on Data Mining, ICDM 2008, pp. 263–272. IEEE (2008)
5. Hong, J., Suh, E.H., Kim, J., et al.: Context-aware system for proactive personalized service based on context history. Expert Syst. Appl. **36**(4), 7448–7457 (2009)
6. Adomavicius, G., Tuzhilin, A.: Context-aware recommender systems. In: Recommender Systems Handbook, pp. 217–253. Springer US (2011)
7. Jamali, M., Ester, M.: A matrix factorization technique with trust propagation for recommendation in social networks. In: Proceedings of the Fourth ACM Conference on Recommender systems, pp. 135–142. ACM (2010)
8. Koren, Y., Bell, R., Volinsky, C.: Matrix factorization techniques for recommender systems. Computer **42**(8), 30–37 (2009)

Bidirectional Analysis Method of Static XSS Defect Detection Technique Based On Database Query Language

Baojiang Cui[1,2]([✉]), Tingting Hou[2], Baolian Long[2], and Lingling Xu[3]

[1] National Engineering Laboratory for Mobile Network Security, Beijing, China
`cuibj@bupt.edu.cn`
[2] School of Computer Science, Beijing University of Posts and Telecommunications, Beijing, China
`houtingtingting@163.com, longbaolian@126.com`
[3] Department of Computer Science, South China University of Technology, Guangzhou, China
`xulingling810710@163.com`

Abstract. Along with the wide use of web application, XSS vulnerability has become one of the most common security problems and caused many serious losses. In this paper, on the basis of database query language technique, we put forward a static analysis method of XSS defect detection of Java web application by analyzing data flow reversely. This method first converts the JSP file to a Servlet file, and then uses the mock test method to generate calls for all Java code automatically for comprehensive analysis. We get the methods where XSS security defect may occur by big data analysis. Originated from the methods where XSS security defect may occur, we analyze the data flow and program semantic reversely to detect XSS defect by judging whether it can be introduced by user input without filter. Moreover, to trace the taint path and to improve the analysis precision, we put forward bidirectional analysis. Originated from the results of the reverse analysis, we analyze the data flow forward to trace the taint path. These two methods have effectively reduced analyzing tasks which are necessary in forward ways. It was proved by experiments on some open source Java web projects, bidirectional and reverse methods not only improved the efficiency of detection, but also improved the detection accuracy for XSS defect.

Keywords: Bidirectional analysis · Web application · JSP file · XSS defect · Static analysis · Reverse analysis

1 Introduction

XSS is a kind of vulnerabilities of Web applications, which is caused by the failure of the application in checking up on user input before returning it to clients web browsers [1]. Without an adequate validation, user input may include malicious scripting code that may be sent to other clients and unexpectedly executed

© Springer-Verlag Berlin Heidelberg 2015
N.T. Nguyen et al. (Eds.): Transactions on CCI XIX, LNCS 9380, pp. 32–44, 2015.
DOI: 10.1007/978-3-662-49017-4_3

by their browsers, thus causing a security exploit [2]. XSS vulnerabilities can be divided into three categories [3]: Reflected XSS (non-persistent type XSS), Stored XSS (persistent type XSS) and DOM Based XSS [4].

Reflected XSS: Reflected XSS occurs when user input is immediately returned by a web application in an error message, search result, or any other response that includes some or all of the input provided by the user as part of the request, without that data being made safe to render in the browser, and without permanently storing the user provided data [5]. This process seems like a reflection, so it is called reflected XSS [6].

Stored XSS: The only difference between stored and reflected XSS is that stored XSS scripting codes are stored on the target server (either database or file system memory, etc.), When the page is requested again, the malicious scripting codes will still be there. The most typical example is the message board XSS [7]: The user submits a message containing XSS scripting codes stored in the database, and when the target user view the message board, those messages will be queried from the database and displayed, the browser execute the XSS code as normal HTML and JS, thus they trigger a XSS attack. Stored XSS attacks are the hardest to be found [8].

DOM XSS: DOM based XSS is a form of XSS where the entire tainted data flow from source to sink takes place in the browser, i.e., the source as well as the sink of the data are both in the DOM, and the data flow never leaves the browser. It can be considered only related to the client [9].

Static analysis technique has been widely researched in source code defect detection. Static analysis can be used for examining codes without modification or execution while completely covering the codes. It is independent of compiler and more convenient and faster than dynamic detection [10]. There are many static analysis methods of defect detection of code, including symbolic execution, theorem proving, type inference, abstract interpretation, rule-based checking and model checking etc. At present, the most mature static Java code scanning tools are Fortify, Findbugs, LAPSE, etc. There are also many tools for cross-site scripting vulnerability detection, such as Paros Proxy [11], XSS-Me [12], etc. However, they all have one or more problems with the analysis efficiency or detection accuracy [13,14].

In this paper, we present our reverse analysis method for static XSS defect detection. We first convert the JSP file to the Servlet file, and then use the mock test method to generate calls for all Java code automatically for the purpose of analyzing comprehensively. In the process of static analysis, we summarize the characters of XSS defect by data mining, then we use Datalog to describe the XSS defect flow, and then use the database query based approach to reversely analyze the method which may cause XSS vulnerability, and confirm the vulnerability by judging whether a related introduction of tainted data can be found. The experimental results show that this method not only improved the efficiency of detection, but also improved the detection accuracy for XSS defect.

2   Reverse XSS Defect Detection Theory

In order to detect XSS defect, in this paper, we use the detection methods based on a database query language Datalog [15]. To describe the defect based on database query language, program information is described as a tuple (D, R, Q), where:

1. D denotes domain, referring to the set of source code attributes $\{x1, x2, x3...\}$. For example, a set D $= \{v, h\}$ where v represents the variable, h represents the heap.
2. R denotes relation. A relation R(x1,..., xn) is a tuple consisting of n-dimensional attributes. Relation R (x1, ..., xn) is true if and only if the statement (x1,..., xn), which is described in relation R, is executed in the program. For example, R = assign (v1, v2) indicates the value of v2 is assigned to the variable v1, and R = vP (v2, h) represents variable v2 points to the heap object h.
3. Q denotes inference rules. A rule Q:

$$Q = Rq(x1, x2, x3, ...) : -R1(x1, x2, x3, ...), ..., Rn(x1, x2, x3, ...) \quad (1)$$

Rule (1) means that when the relation R1,...,Rn are simultaneously true, the relation Rq can be deduced also true. For example,

$$Q1 = vP(v1, h) : -assign(v1, v2), vP(v2, h) \quad (2)$$

Rule (2) indicates that when a variable v2 is assigned to v1, and the variable v2 pointing to heap object h is also executed, we can deduce that the variable v1 point to heap object h is true.

2.1   Domains of XSS Defect Description

In order to describe the XSS flow, we defined the domains in Table 1:

Table 1. XSS defect domains

ID	Set	Description
1	V	program variable
2	H	heap object
3	F	filed variable of an object
4	T	variable type
5	I	call position of a method
6	Z	Number of parameters of the method invocated
7	VC	context properties
8	M (source,derivative, sink)	Invocating method name (name of the methods which introduces, spreads and executes tainted data

The set V, H, F, T, I, Z respectively contain attributes which related to the program, set M represents a collection of methods which related to the program. In the XSS defect description, we divided the set M into source, derivative and sink sets, which respectively include the method names which is associated with XSS tainted datas introduction, spread and execution. In the following we use source method to represent the method which introduces tainted data, derivative method to represent the method which spreads tainted data and sink method to represent the method whose parameter is tainted data.

Through the analysis of the three types of XSS defects we found that, as DOM-based XSS defects mainly exist in JSP code whereas the detection method in this paper is for Java code, we cant detect DOM-based XSS defects. The reason is that the other two kinds of defects are both related to user input, and therefore we take the user input related methods as the source methods. As for storage-based XSS, the user input data may be stored, so we take the methods which are related with database or document reading and writing as a special part of the derivative methods. Reflected and stored XSS defects are related to the implementation of the final output of the tainted data to the JSP page, therefore we take the output methods associated with outputting to the JSP page as sink methods.

2.2   Relations of XSS Defect Detection

In this paper, we defined the relations of XSS defect as described in Table 2: The relations described in Table 2 contain all the operations that may be needed in XSS defect analysis in preparation for reverse analysis rules for XSS defect.

2.3   Reverse Semantic Analysis Rules of XSS Defect Detection

By analysing a method semantically, we judge whether the method will introduce tainted data. If a statement has executed a method that may cause XSS defect, we mark the related message as tainted data, and its possible source path is illustrated as Fig. 2. Where status A indicates the executed method may lead to XSS defects, status B indicates tainted data propagation, and status C represents the introduction of tainted data.

Path A-B-C Means the parameter of the sink method comes from a derivative method, and the parameter of the derivative method comes from the source method, so this can define a XSS defect.

Path $A^{\iota} - B^{\iota} - C^{\iota}$ Means the parameter of the sink method comes from a non-derivative method, but the parameter of the non-derivative method comes from the source method, so this cant refer to a XSS defect.

Path $A^{\iota\iota} - B^{\iota\iota} - C^{\iota\iota}$ Means the parameter of the sink method comes from a derivative method, but the parameter of the derivative method doesnt come from a source method, so this cant refer to a XSS defect.

From the above analysis, in order to detect a XSS defect we only need to consider path, namely the parameter of sink method originating from a derivative

Table 2. XSS defect relations

ID	Relation	Description
1	MS (M0, S0)	Methods set M0 belongs to source methods set S0, namely M0 ∈ S0
2	MK (M0, K0)	Methods set M0 belongs to sink methods set K0, namely M0 ∈ K0
3	MD (M0, D0)	Methods set M0 belongs to derivative methods set D0, namely M0 ∈ D0
4	ME (M0, E0)	Methods set M0 belongs to special derivative methods set E0, namely M0 ∈ E0
5	S (V0, F0, V1)	Save operation, the value of the variable V1 is assigned to the field variable F0 of variable V0, namely V0.F = V1
6	L (V0, F0, V1)	Fetch operation, variable V1 fetch the value of the field variable F0 of variable V0, namely V1 = V0.F
7	IEcs (VC1,I0,VC0,M0)	Function call, at call site I0 with the context VC1, called a method M0 with context VC0
8	Iret (I0, V0)	Return value of call site, the return value of call site I0 is V0
9	cvP (VC0, V0, H0)	New operation, V0 with context VC0 point to object H0namely V0 = new Object()
10	cA (VC0, V0, VC1, V1)	Assignment operation, the value of variable V1 with context VC1 is assigned to the variable V0 with context VC0, namely V0 = V1
11	actual (I0, Z0, V1)	Actual parameter of a method, the Z0th actual parameter of method called at call site I0 is V1
12	sinked (VC0, V0,H0,I1,M1)	Execution sink method, domain variable H0 of variable V0 with context VC0 is tainted datum; tainted data come from the sink method M1 executed at call site I1
13	sinkedhP (H0,F0, H1,I1,M1)	Tainted data assignment, field F0 of object H0 point to tainted data H1, tainted data come from the sink method executed at call site I1
14	derics (VC0,V0,VC1,V1)	Tainted data spared, tainted variable v1 with context VC1 is propagated to variable V0 with context VC0, namely V0 = V1

method, and during the propagation the tainted data not being operated by any filtering method (non-derivative method), and the tainted data originating from source method. After analyzing the above rule, we got the reverse analysis rules of XSS defect as following:

(1)

$$\left.\begin{array}{r} IEcs(vc1, i, \_, m) \\ MK(m, \_) \\ actual(i, \_, v) \\ cvp(vc, v, h) \\ vc = vc1 \end{array}\right\} - > sinked(vc, v, h, i, m)$$

If call position i with context vc1 called sink method m, the actual parameter of call position i is v, at the context $v-> h, vc = vc1$, then we can deduce that v with context vc is a tainted datum.

(2)

$$\left.\begin{array}{r} sinked(vc, v, h, i, m) \\ cA(vc, v, vc1, v1) \end{array}\right\} - > sinked(vc1, v1, h, i, m)$$

If v with context vc is a tainted datum, and $v = v1$, then we can deduce that v1 with context vc1 is also tainted.

(3)

$$\left.\begin{array}{r} sinked(vc1, v2, h2, i, m) \\ L(v1, f, v2) \\ cvP(vc1, v1, h1) \end{array}\right\} - > sinkedhP(h1, f, h2, i, m)$$

If v2 with context vc1 is tainted, and $v2 = v1.f$, at context vc1 $v1-> h1$, then we can deduce that $h1.f-> h2$ is also tainted.

(4)

$$\left.\begin{array}{r} sinkedhP(h1, f, h2, i, m) \\ S(v1, f, v2) \\ cvP(vc1, v1, h1) \end{array}\right\} - > sinked(vc1, v2, h2, i, m)$$

If $h1.f-> h2$ is tainted, and $v1.f = v2$, at context vc1 $v1-> h1$, then we can deduce that v2 with context vc1 is also tainted.

(5)

$$\left.\begin{array}{r} IEcs(vc1, i, vc2, m) \\ MD(m, \_) \\ actual(i, \_, v2) \\ Iret(i, v1) \end{array}\right\} - > derics(vc1, v1, vc2, v2)$$

If call position i with context vc1 called derivative method m with context vc2, the actual parameter of method m is v2, and the return value is v1, then we can deduce that v2 with context vc2 is propagated to v1 with context vc1.

(6)

$$\left.\begin{array}{r} IEcs(vc1, i, vc2, m) \\ ME(m, \_) \\ actual(i, 0, v2) \\ Iret(i, v1) \end{array}\right\} - > derics(vc1, v1, vc2, v2)$$

If call position i with context vc1 called a special derivative method m with context vc2, the actual parameter of call position i is v2, the return value is v1, then we can deduce that v2 with context vc2 is spread to v1 with context vc1.

(7)

$$\left.\begin{array}{r} derics(vc1, v1, vc2, v2) \\ sinked(vc2, v2, \_, i, m) \\ cvP(vc1, v1, h1) \end{array}\right\} -> sinked(vc1, v1, h1, i, m)$$

If v2 with context vc2 is tainted, and it is propagated to v1 with context vc1, at context vc1 $v1-> h1$, then we can deduce that v1 is also tainted.

(8)

$$\left.\begin{array}{r} sinked(vc1, v, h, i1, m1) \\ Iret(i, v) \\ MS(m, \_) \\ IEcs(vc, i, \_, m) \\ vc = vc \end{array}\right\} -> source(i, m, v, h, i1, m1)$$

If v with context vc1 is tainted data, call position I with context vc called source method m, return value is v, and $vc = vc1$, then we can deduce that call position I has introduced tainted data.

Compared with the forward data flow analysis, since the number of the methods which may cause XSS defect is much less than those which may introduce tainted data, the reverse analysis method can reduce the number of methods which need to be analyzed to significantly improve the static analysis efficiency. At the same time, we use the database queries based approach, so only if the sink method set K, derivative method D, special derivative method E, and source method S are complete and reliable, we can guarantee detection results with high accuracy.

2.4   Bidirectional Taint Propagation Analysis Approach

Through the reverse analysis above, we get the static analysis results with high precision. However, we can not get the taint propagation path and the precision can also be improved. So we put forward Bidirectional Taint Propagation Analysis Approach to tracking the taint propagation path and improve the precision. Bidirectional analysis is based on the results of reverse semantic analysis. The process of bidirectional analysis is as below:

(1) Analyze the data flow reversely, and get the source() and sinked() set which both save the sinking call site $I_s ink$ and method $M_s ink$.
(2) Create the subset of sinked() set according to every $I_s ink$ and $M_s ink$ in source() set.
(3) Get the taint propagation path from source() to sinking() according to the subset of sinked() above and the derics().

After analyzing the rules above, we got the bidirectional analysis rules of XSS defect as following. Among them, a tuple $< i, m >$ is defined as a propagation node in the taint propagation path.

(1)

$$source(vc, i, m, v, h, i1, m1) - > taintedforward(vc, v, h, i, m, i1, m1, i, m)$$

Set the initial value of taintedforward() using source(). Vc,v and h are the basic information of taint object to be saved. $< i, m >$ is the source node in the taint propagation path. The first $< i, m >$ mean the last propagation node. The second $< i, m >$ mean the current propagation node. $< i1, m1 >$ means the last propagation node where the taint data is sink.

(2)

$$\left. \begin{array}{l} taintedforward(\_, \_, h, i, m, i1, m1, i, m) \\ sinked(vc1, v1, h, i1, m1) \end{array} \right\} - > taintedforward(vc1, v1, h, i, m, i1, m1, i2, m2)$$

If the heap object h is in the taintforward() set, we can deduce that any variable v1 with context vc1 is in the taintforward() set if it points to h. Other information such as i and m stays the same.

(3)

$$\left. \begin{array}{l} derics(vc, v1, v2, i, m) \\ taintedforward(vc, v1, \_, \_, \_, i1, m1, i2, m2) \\ cvP(vc, v2, h2) \end{array} \right\} - > taintedforward(vc, v2, h2, i2, m2, i1, m1, i, m)$$

When the tainted data is propagated from variable v2 to v1 through node $< i, m >$, the node $< i, m >$ will be saved in the taintedforward() set as the current propagation node. And $< i2, m2 >$ is saved as the last propagation node.

(4)

$$\left. \begin{array}{l} taintedforward(vc, v2, h, \_, \_, i1, m1, i, m) \\ Iret(i1, v2) \end{array} \right\} - > taintedforward(vc, v2, h, i, m, i1, m1, i1, m1)$$

When v2 is the return value of the sinking node $< i1, m1 >$, i1 and m1 is saved as the current node. The propagation process is end when it reaches the sinking node. Then we can deduce the taint propagation path from the source node to the sinking node by writing an additional program.

From the above, we can get the taint propagation path. Moreover, in the additional program, we have filtered the reduplicative results and some wrong results. For example, JspWriterImpl.write() and JspWriterImpl.print() are both sinking method, but in some case JspWriterImpl.print() will invoke JspWriterImpl.write(). So if the call site I of method JspWriterImpl.write() is JspWriterImpl.print(), we can filter this path. This happens often in static analysis, so the precision is improved very much.

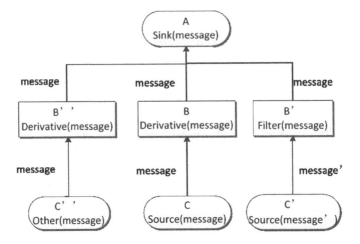

Fig. 1. XSS defect detection process

3   Implementation of Reverse Static Detection on XSS Defect

After analyzing the java application source code, we designed the process of XSS defect detection as Fig. 1:

3.1   Pre-processing Program

The system uses the Java web application source code as input, including JSP files, Java Servlet files and web.xml file which contains call information. Since reflected XSS defects may exist in JSP file, we translate JSP files to Java Servlet files first, so that we can analyze them. Then use the mock testing method to utilize web.xml file to generate calls for all Java source code, so that the static analysis can cover all Java code while reducing false negatives [16–18].

3.2   Call Information Analysis of the Source Code

Use the Java code analyzer to generate the program call information. It includes the set of all the method, invoke sites, fields, string and types. It also includes the relation of assignment, method invoke, pointer, etc. All the relations are generated from the entrance method of mock test generated, and are saved in relational database form.

3.3   Program Information Analysis

Program Information Analysis is based on the relational database queries. It uses the database query based analyzer called bddbddb [19], and applies the

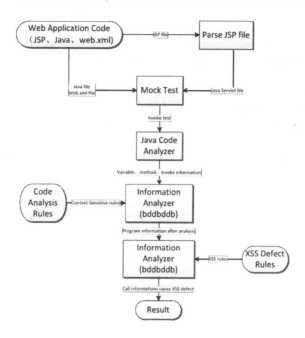

Fig. 2. Tainted data reverse analysis path

context-sensitive analysis [20] rules to analyze the program information, which was generated in the 3.2 analysis, to simplify the program information. The output of program information analysis is the alias information of variable in different application context [21,22].

3.4 Defect Rules Matching Analysis

Apply the reverse XSS defect analysis rules to analyze the information which was generated in 3.3, and output the XSS defects analysis results. The reverse XSS defect analysis rules are listed in Chap. 3, we transfer the rules into relational database query language called Datalog [19].

4 The Results and Analysis of Experiments

4.1 Comparing Results of Forward Analysis, Reverse Analysis and Bidirectional Analysis

In order to compare the efficiency of the proposed two XSS defect detection methods above with forward XSS defect detection methods, we tested six open source Java web projects by using forward [23], reverse and bidirectional detection method respectively. To demonstrate the effectiveness of this method, we have compared the results with business software called Fortify. Fortify is a commercial software of source code analysis. The number of analyzed tainted data number, path number, analysis time and the detected XSS defect number are given in Fig. 3 (Where T represents tainted data number, P represents path number, AT represents analysis time and R represents XSS defects reported/false positive).

Figure 3 summarizes the results for the four different analysis versions. The first part of the table shows the tainted objects, analysis time and false positive of forward analysis. The second part of the table shows the tainted objects, analysis time and false positive of reverse analysis. The third part of the table shows the tainted objects, path numbers, analysis time and false positive of reverse analysis. The last part of the table shows the analysis results of Fortify. Figure 4 provides a graphical representation of the number of tainted objects for different analysis variations. Below we summarize our observations.

Test project	forward analysis			reverse analysis			bidirectional analysis				Fortify analysis			
	T	AT	R	T	AT	R	T	P	AT	R	T	P	AT	R
WebGoat-5.4	805	23.7 min	32/1 1	398	16.7 min	35/11	214	321	48.6 min	16/11	1920	4203	18.9 min	289/2 6
blojsom-quickrstart-3.3b	457	16.5 min	6/4	420	12.8 min	7/3	352	402	32.6 min	5/4	682	2469	13.4 min	85/6
blueblog-0.9	61	8.4 min	8/2	37	6.5 min	8/1	35	127	17.1 min	2/1	99	236	6 min	15/2
pebble-2.6.4	516	19.8 min	13/6	385	18.7 min	14/6	193	264	40.3 min	9/6	914	1058	13.4 min	67/6
jboard-0.30	254	14.7 min	18/7	196	12.4 min	18/5	174	410	28.4 min	9/7	327	541	9.2 min	39/9
geoserver-2.5-beta	769	20.3 min	11/5	521	18.5 min	12/3	329	567	42 min	6/5	1032	3852	13.4 min	37/6
Average	477	17.2 min	14.7 /7	326 .2	14.3 min	15.7/ 4.8	216. 2	348 .5	34.8 min	7.8/6. 2	829	2059. 8	12.4 min	88.7/ 9.2

Fig. 3. Summary of data on the number of tainted objects, paths, analysis time and false positive for each analysis version.

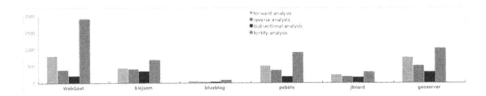

Fig. 4. Comparison of the number of tainted objects for each version of the analysis.

Comparing reverse analysis with forward analysis, we can see that the number of the analyzed tainted data number and analysis time by reverse analysis method are all less than those of forward analysis method, while the precision is not less than the forward analysis method, which can prove that the reverse XSS defect analysis methods has ensured the XSS defect detection precision and improved the scanning efficiency.

Comparing reverse analysis with Bidirectional analysis, we can see that Bidirectional analysis get less tainted data number and higher precision. Moreover, bidirectional analysis also output the taint propagation path. However, as we can see, the high precision is at the cost of analysis time. The analysis time of bidirectional analysis is more than any other analysis method.

Comparing bidirectional analysis with Fortify analysis, it should be noted that the taint analysis time often decrease as the analysis precision increases. Wrong results take too much time to verify, so, we believe that this is acceptable given the quality of the results the analysis produces.

5   Conclusion and Future Work

This paper has represented how XSS in Java applications can be formulated as instances of the general tainted object propagation problem, which involves finding all sink objects derivable from source objects via a set of given derivation rules. We have put forward two semantic analysis methods in the paper to detect XSS in Java source code.

Reverse analysis analyzes the taint data from the sinking site to the source site. It traces the propagation of taint data reversely. Compared with forward analysis, reverse analysis would save analysis time because the sinking nodes are less than source node of tainted data.

Bidirectional analysis is based on reverse analysis. It analyzes the taint data bidirectionaly, which means analyze the source code reversely and then forward. The results are more precise and the taint data propagation path is also produced as appendage. We have compared the analysis results with Fortify. It turns out that the bidirectional analysis get lower false positive rate.

Acknowledgments. This work was supported by National Natural Science? Foundation of China (No.61170268, 61100047, 61272493)

References

1. Yawen, W.: Defect model based software testing technology. Beijing Univ. Posts Telecommun. (2009)
2. Di Lucca, G.A., Fasolino, A.R., Mastoianni, M., Tramontana, P.: Identifying cross site scripting vulnerabilities in Web applications. In: 26th Annual International Telecommunications Energy Conference, INTELEC 2004, pp. 71–80, 11 September 2004
3. Open Web Application Security Project. Types of Cross-Site. October 2013 Scripting (2013). https://www.owasp.org/index.php/Types_of_Cross-Site_Scripting
4. Zhong Chenming, X.S.: Web Front-endReveal Hacking Techniques. Electronic Industry Press, Beijing (2013)
5. Martin, M., Lam, M.S.: Automatic generation of XSS and SQL injection attacks with goal-directed model checking. In: Proceedings of the 17th Conference on Security Symposium, (pp. 31–43). USENIX Association (2008)
6. Bisht, P., Venkatakrishnan, V.N.: XSS-GUARD: precise dynamic prevention of cross-site scripting attacks. In: Zamboni, D. (ed.) DIMVA 2008. LNCS, vol. 5137, pp. 23–43. Springer, Heidelberg (2008)

7. Fonseca, J., Vieira, M., Madeira, H.: Testing and comparing Web vulnerability scanning tools for SQL injection and XSS attacks. In: 13th Pacific Rim International Symposium on Dependable Computing, 2007, PRDC 2007, pp. 365–372. IEEE (2007)
8. Wurzinger, P., Platzer, C., Ludl, C., Kirda, E., Kruegel, C.: SWAP: Mitigating XSS attacks using a reverse proxy. In: Proceedings of the 2009 ICSE Workshop on Software Engineering for Secure Systems, pp. 33–39. IEEE Computer Society (2009)
9. Klein, A.: DOM based cross site scripting or XSS of the third kind. Web Application Security Consortium, Articles, 4 (2005)
10. Fonseca, J., Vieira, M., Madeira, H.: Testing and comparing Web vulnerability scanning tools for SQL injection and XSS attacks. In: 13th Pacific Rim International Symposium on Dependable Computing, 2007, PRDC 2007, pp. 365–372. IEEE (2007)
11. Paros, Y.: Paros Proxy [DB/OL] (2006). http://sourceforge.net/projects/paros/
12. Mozilla. XSS-Me [DB/OL] (2012). http://labs.securitycompass.com/exploit-me/xss-me/
13. Vogt, P., Nentwich, F., Jovanovic, N., Kirda, E., Kruegel, C., Vigna, G.: Cross Site scripting prevention with dynamic data tainting and static analysis. In: NDSS (2007)
14. Wassermann, G., Su, Z.: Static detection of cross-site scripting vulnerabilities. In: ACM/IEEE 30th International Conference on Software Engineering, 2008, ICSE 2008, pp. 171–180. IEEE (2008)
15. Benjamin Livshits, V., Lam, M.S.: Finding security vulnerabilities in java applications with static analysis. In: USENLX Technology Symposiu (2005)
16. Kirkegaard, C., Møller, A.: Static analysis for java servlets and JSP. In: Yi, K. (ed.) SAS 2006. LNCS, vol. 4134, pp. 336–352. Springer, Heidelberg (2006)
17. Chess, B., West, J.: Secure Programming with Static Analysis. Pearson Education, USA (2007)
18. Haviv, Y.A., Tripp, O., Weisman, O.U.S.: Patent No. 8,726,245. Washington, DC: U.S. Patent and Trademark Office (2014)
19. Whaley, J., Dzintars, A., et al.: Using datalog with binary decision diagrams for program analysis. In: Third Asian Symposium (2005)
20. Whaley, J., Lam, M.S.: Cloning-based context-sensitive pointer alias analysis using binary decision diagrams. In: Proceedings of the ACM SIGPLAN 2004 Conference on Programming Language Design and Implementation (PLDI) (2004)
21. Whaley, J., Lam, M.S.: Cloning-based context-sensitive pointer alias analysis using binary decision diagrams. In: ACM SIGPLAN Notices, vol. 39, no. 6, pp. 131–144. ACM (2004)
22. Tripp, O., Pistoia, M., Cousot, P., Cousot, R., Guarnieri, S.: ANDROMEDA: accurate and scalable security analysis of web applications. In: Cortellessa, V., Varró, D. (eds.) FASE 2013 (ETAPS 2013). LNCS, vol. 7793, pp. 210–225. Springer, Heidelberg (2013)
23. Schneier, B.: Applied Cryptography: Protocols, Algorithms, and Source Code. Wiley, New York (2007)

A Multilevel Security Model for Search Engine Over Integrated Data

Gansen Zhao[1]([✉]), Kaijun Chen[1], Haoxiang Tan[1], Xinming Wang[1,2],
Meiying You[1], Jing Xiao[1], and Feng Zhang[3]

[1] School of Computer Science, South China Normal University, Guangzhou, China
`gzhao@scnu.edu.cn, wangxm35@mail.sysu.edu.cn`
[2] School of Software, Sun Yat-sen University, Guangzhou, China
[3] School of Computer Science, China University of Geosciences, Wuhan, China

Abstract. Data has become a valuable asset. Extensive work has been put on how to make the best use of data. One of the trends is to open and share data, and to integrate multiple data sources for specific usage, such as searching over multiple sources of data. Integrating multiple sources of data incurs the issue of data security, where different sources of data may have different access control policies. This work investigates the issue of access control over multi data sources when they are integrated together in the scenario of searching over these data. We propose a model to integrate multiple security policies while data are integrated to ensure all data access respects the original data's access control policies. The proposed model allows the merging of policies and also tackles the issue of policy conflicts. Theoretical analysis has been conducted, which suggests that the proposed model is correct in terms of retaining all original the access control policies and ensure the confidentiality of all data.

Keywords: MLS · Lattice · Multiple data source · Data integration · Data confidentiality

1 Introduction

The increasing volume of data is attracting attentions and curiosities on making the best use of the data and get the best value out of it. One of the development trends is to open data for others to use, and to integrate data from different sources together. Hence, more comprehensive data sets can be constructed and could potentially support more sophisticated data analysis, data mining as well as decision makings.

One of the issues that have to deal with is the confidentiality of the data that are integrated and accessed. When data from multiple sources are integrated and searched, search engines must ensure that the returned search results do not contain data that are not accessible to users. For example, a user u has access permissions to the data set D_A but not the data set D_B. In the integration process, a data set D_C is generated based on D_A and D_B. In other terms, D_C

© Springer-Verlag Berlin Heidelberg 2015
N.T. Nguyen et al. (Eds.): Transactions on CCI XIX, LNCS 9380, pp. 45–68, 2015.
DOI: 10.1007/978-3-662-49017-4_4

contains information from both D_A and D_B. When the search engine responses to a query q issued by the user u, the search result should filter those data that are related to data from D_B. This is because if data related to D_C is presented to u, u could potentially obtain information from D_B.

Secure search has developed for several years, such as Google Search Appliance. These systems enable secure search using access control lists, but they are not adequate for our scenario in at least the following two aspects. Firstly, most search services enforce access control using ACLs, which fail to control information flow, making it difficult to prevent information leakage. Secondly, most solutions do not consider the policy transformation after data integration. Different data sources may governed by their own policies, which must be respected at all time.

This work assumes that all data sources are governed by Bell-LaPadula (BLP) security policies. The mandatory access control of BLP policies is based on the multi level security (MLS) model, which is a system model that assigns security clearance to subjects and security levels to objects for the purpose of controlling system information flow. The discretionary access control of BLP policies is based on access control matrices.

This work proposes a method to combine multiple BLP policies together while the corresponding data sources are integrated together. The combining of multiple BLP polices will generate a new BLP policy, which could be used to govern the access to the new integrated data. The new BLP policy ensures all access to the integrated data be consistent with the original security policies of the corresponding data sources.

The proposed method has two main parts: lattice integration and matrix integration. Lattice integration is implemented as the merging of Hasse graphs, which has the initial phase, the conflict management phase and the simplification phase, while matrix integration is implemented as combining of two access matrices.

Theoretical security analysis has been conducted, whose result suggests that the proposed method is able to generate a new security policy that provides the appropriate security access control to the integrated data.

Our contributions can be summarized as follows. Firstly, we formalize the system of searching over integrated data from multiple data sources of different BLP security policies. Secondly, we propose a method to combine BLP policies according to the process of data integration. Thirdly, we perform a theoretical security analysis of the proposed method and show that the method is capable of generating the right security policy for access control.

The rest of this paper is organized as follows. Section 2 presents a comprehensive review over existing work. Section 3 shows a scenario of searching over integrated data. An architecture to enable secure search is proposed. The mathematical model is also developed in this section. Section 4 proposes a method for access policy integration to generate new security policies according to the data integration process. Section 5 conducts a security analysis on the proposed method. Section 6 concludes this research and also identifies future work.

2   Related Work

2.1   Data Integration and Secure Search

Data integration is a process combining multiple data sources and providing a unified interface to access them. Generally, [9,10,18] describe an integration system as a tripe: global-as-view (GAV), local-as-view (LAV) and mapping between them. These approaches separate a system into three components and create a mediator to handle the data heterogeneity and query processing.

Motivated by data sharing and opening, data integration is increasingly used in cloud and data-intensive applications. Confidentiality of data must be maintained after integration [1,7,17,20]. The main challenges faced in protecting privacy over integrated data is the need of query across multiple data sources owned by multiple organizations having different security paradigms [9,11]. This combination may blur the data boundary and impact the access privileges.

There are several industrial and academic search systems to support privacy protection for searching over integrated data. A variety of commercial systems have been deployed, such as Google Search Appliance (GSA). GSA [8] applies two types of ACLs: Per-URL ACLs and Policy ACLs. The former is to assign authorization rules to each document in the index, while the latter one protects URL patterns rather than individual URL. Through Access Connector, the user can hold their credentials to various data repositories. In the fields of academic research, most research work are concerning about searching over encrypted data. To address the efficiency issue of searching over encrypted data due to the overhead of decryption, several works [4,12,22] use coordinate matching and inner product similarity to improve efficiency and enhance security.

2.2   Multi Level Security

To achieve the mandatory control of information flows, many research on data confidentiality use Multi Level Security (MLS) or Bell-LaPadula (BLP) models. In Multilevel Security Systems [13], users and data are labeled by tags or attributes which are respectively called users' clearance level and the data's classification level. If and only if a clearance level of the users meet a certain requirements with the data's classification level, such as higher than the data's class level then access can be allowed.

There has been a large body of work on access control to implement MLS model. To partitioning workflows for processing the medical data under the federated clouds, Watson [24] describes a novel model to split the services into different service states and give the security level to each service states. By defining the security data transformation rules and the cost model for execution, the security of the model can be proved. To handle less restrictive and general distributed systems or networks, Watson et al. [23] would consider the platform and network as an object, which are tagged by security levels. So when selecting a service to run, data would check not only the service states, but also the platform and network which would be chosen.

To meet the needs of providing flexible access control based on MLS, some researches argument factors on the security level. Su et al. [19] describe the idea of putting environmental and temporal states of subjects and objects into the access control model, when determining whether subjects are allowed to access objects. It can meet the need of flexible and fine-grain access control. Xue et al. [26] propose a model that the subject security level can dynamically change when accessing the sensitive data. By giving the state transition function rules, it is proved to be secure while the user changes the current security level. Besides, another disadvantage of MLS model is that it is a labor-consuming job for administrators to assign the labels to data or to users. Thorleuchter et al. [21] use the latent semantic indexing to assign the textual information to security labels. By text classification methodology, this model can successfully assign the labels to information concerning the semantic of their content. This can reduce the complexity of security managements of administrators.

2.3   Integration of Security Policies

Policy integration approaches are to combine more than one policies and generate a policy compliant with the original security requirements [18]. Users would send their queries through a unique entry point, so-called mediator, and they can get query results from different data sources [9]. But, the access policies specified in different data sources may be in conflict with each other. It is difficult to manage the policy conflict information in multi access control policies [14]. Therefore, federated management is introduced to tackle the conflicts when combing the policies from their original data sources [6].

A number of papers have proposed different ways to deal with policy conflicts. One of the proposals is to use mathematical logic calculation. Rao [15,16] introduces an algebra for composing privacy policies and proposes a framework using algebra for the fine-grained integration of policies expressed in XACML [3]. Using XACML to describe an access policy, the combing of multiple access policies is to calculate mathematical logic of each policy. Because of incomplete property or information provide by users, the final results of mathematical logic calculations may be unknown (NA). This will damage the availability of the system to some extent.

Another method of dealing with conflicts is to rewrite the query sentences or to generate a global mapping schema. For example, a query Q can be transformed to Q' which brings about the result compliant with the policy of the original data source. Hu [11] describes semantics-enabled policies. By means of ontology mapping and merging, queries would be rewritten which mapping names of class entities and properties to the local queries. Besides, local policy can be mapped into the global in a schema. Cruz et al. [5] combine local policies stored in XML schemes and transform into a global RDF schemes. When two policies are merged, local schemes are transformed into RDF schemes before being transformed into global RDF schemes. Alodib [2] presents a method of utilizing the Access Control Policy Service in a service oriented environment, which merges the information of requester, provider, and policies in different web services by The Web Services Description Language (WSDL).

3 Scenarios and Systems

3.1 Search Systems

Searching over integrated data works as follows. There are multiple data sources, each of which contains several data sets. These data sources are integrated together, generating a new data base. Search queries are executed over the integrated data base. Search results are then presented to users. Figure 1 demonstrates a simplified model of a search engine over integrated data. Some key elements are involved in the model, including Users, the Search Engine, the Integrated Database, Data Sources, Data Sets and Data Items, etc.

- Users. End users of search services, who submit search queries over the integrated data base, and will access all returned results.
- Keywords. Terms are used to represent users' queries.
- Search Engine. The search engine is responsible for responding to the queries submitted by users and executing the queries over the integrated data base, after which it will present the search results to the users.
- Integrated Database. An integrated data base integrates multiple data sets.
- Data Sources. Data sources are provided by the corresponding parties and the data are open to subjects under appropriate security policies.
- Data Sets. Each data set is a set of data that could be of interest to users, and the data are protected by security policies.
- Data Items. Data items are the minimum data unit in search system.
- Results. The data generated by the execution of users' queries over the integrated data.

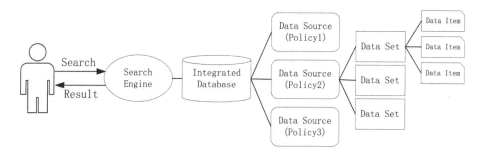

Fig. 1. Search system

One of the scenarios is searching over taxation data. The Tax Bureau is a hierarchical organization. Each department of the Tax Bureau is responsible for different duties and can access different data, with each department being a individual data source. When taxation data from different departments are integrated together, queries can be executed over the complete taxation data. One of the security requirements in this scenario is to regulate potential information flows, making sure that no breach of confidentiality occurs.

3.2 The Integration Architecture

The search system combines data sets from multiple data sources which are governed by their own security policies. Therefore, the process of integration should involve two key steps:

1. Data integration. A series of data sets can be combined using set operations, such as $\bigcup$, $\times$ or other calculations, to generate an integrated data set.
2. Policy integration. After integrating policies, the conflict in the merged policy with the original policies needs to be resolved. A method to tackle the conflicts should be invoked.

With these two steps, we propose an architecture to meet the demand of the search scenario over integrated data. Figure 2 shows the proposed search architecture.

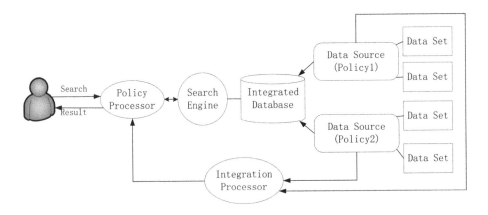

Fig. 2. System architecture

In this architecture, we add two components. (1) The Integration Processor is responsible for integrating policies in the original data sources and generate a new policy. (2) The Policy Processor is responsible for verifying users and data sets by the merged policy and filtering the confidential results for the users.

3.3 Formal Model

Symbols and Terminology. Follows are the key elements of the proposed system.

- S : Subjects. The set of subjects in a search system, with each subject representing an end user.
- O : Objects. The set of objects in a search system, with each object being a data set.
- DS : Data Sources. The set of data sources, with each data source containing multiple data sets.
- C : Classifications. The set of classifications that are developed to represent confidentiality levels.
- K : Categories. K is the set of categories. Each category indicates a relevant domain or topic.
- L : Security Levels. The set of security levels, defined as $L = C \times K$.
- f : Level Function. It is a mapping function that maps a subject or a object to a security level.
- M : Access Matrix. $M \subseteq S \times O$, representing each subject's access rights to each object. $M(s, o) = True \in M$ means that subject s has the access right to object o. Note that only reading access is considered in this work.
- A mapping function s_s, which is used to mapping data sets to their original data sources. Let data set o_i be from data source ds_j, it can be noted as $ds_j = s\_s(o_i)$.
- A mapping function s_d, which is used for mapping result sets to their original data sets. Let a result set rs_i be from the data sets o_j, it can be noted as $o_j = s\_d(rs_i)$.

Security Policy. Assuming that each data source is protected by a BLP security policy. Each subject s is assigned a security clearance $sc \in L$ and each object o is assigned a security level $sl \in L$. Let the BLP security policy (f_i, LTC_i, M_i) be the BLP security policy for a given data source i.

- f_i is the level function.
- LTC_i is the security level lattice defined as $LTC_i = (L_i, \leq_i)$ where L_i be the security level set of Ri and $\leq_i \subseteq L_i \times L_i$ be a partial order. Assume that $l_1 = (c_1, k_1)$ and $l_2 = (c_2, k_2)$, $l_1 \leq_i l_2$ iff $c_1 \leq c_2$ and $k_1 \subseteq k_2$.
- M_i is the access matrix. $(s_i, o_j) = True$ represents $s_i \in S$ has privilege to read $o_j \in O$.

Definition of Data Integration. Let $o_i \in O$ and $o_j \in O$, o_i and o_j is integrated into m_{ij}, denoted as

$$m_{ij} = Int(o_i, o_j)$$

The Int function combines two sets to generate a new set by either using set operations such as $\bigcup$, $\times$, etc., or other calculations. For the process of integrating more than two data sets together, it is defined as follows.

$$m_n = Int(\sum_{i=1}^{n} o_i) = Int(m_{n-1}, o_n)$$

$$m_{n-1} = Int(\sum_{i=1}^{n-1} o_i).$$

Definition of Search. We define a function *search* to describe the process of query. This function has three parameters, u, the given user, $kwds$, the keywords used to query, and m, the integrated data base from different data sets. RS is the result set for the query. This can be formalized as below.

$$RS = search(u, kwds, m)$$

where m is the integrated data base, generated as

$$m = Int(\sum_{i=1}^{n} o_i).$$

Security Requirement. For a given user u, the set of data sets O^i from data source ds_i can be separated into O_{allow}^i and O_{deny}^i, where O_{allow}^i is the data sets accessible for u, O_{deny}^i is the data sets confidential for u. They satisfy both $O^i = O_{allow}^i \cup O_{deny}^i$ and $O_{allow}^i \cap O_{deny}^i = \emptyset$. Based on the policy P_i, O_{allow}^i should satisfy:

$$\forall x(x \in O_{allow}^i) \rightarrow f_i(u) \geq f_i(x) \wedge M_i(u, x) = True$$

Hence,

$$O_{deny}^i = O^i - O_{allow}^i$$
$$= O^i - \{x | (f_i(u) \geq f_i(x) \wedge M_i(u, x) = True)\}$$

Therefore, all the unaccessible data sets to user u can be noted as

$$O_{deny} = \bigcup O_{deny}^i$$

With the integrating process, the integrated data base is generated as $m_n = Int(\sum_{i=1}^{n} o_i)$. Through the mapping function s_s, the original data sources of m_n can be noted as $DS = s_s(m_n) = \{ds_1, ds_2, \ldots ds_k\}$, where k is the number of the data sources. Let P_i be the policy governing data source ds_i. Let P_G be the integrated policy of m_n, hence, it can be noted as:

$$P_G = \sum_{i=1}^{k} Int_p(P_i)$$

where Int_p is a function to integrate policies and generate a new merged policy.

In the integrated search engine, u uses $kwds$ searching on m and gets the results

$$RS = search(u, kwds, m)$$

By using mapping function s_d, the original data sets of RS can be noted:

$$O_{RS} = s\_d(RS) = \{o_{r1}, o_{r2}, \ldots, o_{rn}\}$$

To ensure the confidentiality of the data while searching over integrated data, the returned results should not contain any results from those data sets that are not allowed to be accessed by the user u.

$$\forall o(o \in O_{RS} \rightarrow o \notin O_{deny}).$$

4   Access Policy Integration

Every data source is governed by a BLP security policy. Let an access control policy $P_i = (f_i, LTC_i, M_i)$, where i represents the ith data source. When multiple data sets are merged together, they are to form an integrated data base. Due to the difference existing on data sources, an integrated policy $P_G = (f_G, LTC_G, M_G)$, compliant with the original policies, needs to be constructed. The integration process of security policies consists of two parts, the Lattice integration and the Access Matrix integration.

The Lattice Integration is a process of merging different Lattice to produce a new lattice. The access matrix integration is to expand matrix columns and rows and fill in the access authorities.

4.1   Integration Sequence

In this paper, we consider that the integration method is only to merge two policies in each round.

Let P_i and P_j be two policies, P_i and P_j is integrated into P_{ij}, denoted as

$$P_{ij} = Int_p(P_i, P_j)$$

The Int_p function combines two policies of data sources to generate a new merged policy.

When there is a need to merge more than two policies, it can use this method recursively. For the process of integrating more than two policies together, it is defined as follows.

$$P_n = Int_p(\sum_{i=1}^{n} P_i) = Int_p(P_{n-1}, P_n)$$

where

$$P_{n-1} = Int_p(\sum_{i=1}^{n-1} P_i)$$

For example, assuming that we have N policies of data sources, during the integration process, first two policies are merged into a temporary P'. And then

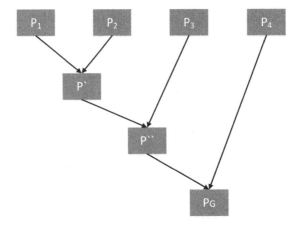

Fig. 3. Integration sequence

choose one in the left $N - 2$ data sources to merge with P'. Recursively using $N - 1$ times, we would obtain the final integrated policy P_G.

In Fig. 3, choose P_1 and P_2 at first, and form a new policy P'. And then merge P_3 with P', which produce another new policy P''. Finally, after integrating with P_4, the final policy P_G is generated.

4.2   Lattice Integration

In order theory, a Hasse graph [25] is a type of mathematical diagram used to represent a finite partially ordered set, in the form of a drawing of its transitive reduction. For a partially ordered set $< S, \leq >$, each element in S can be considered as a vertex, and we can draw a directed line downward from $s1$ to $s2$ if and only if $s1, s2 \in S$ and $s1 \leq s2$.

As lattices are particular type of partially ordered sets, lattices can be represented in Hasse graphs. The merging of lattices can be considered as integrating two Hasse graphs into one. The merging process has three phases, the initial phases, the conflict management phase and the simplification phases. The initial phase is to add the relation lines between the two original Hasse graphs. After adding these lines, the graph may have conflicts with the original policies so that we need to detect and tackle these conflicts in the conflict management phase. Lastly, we could simplify the graph to remove redundant lines.

The Initial Phase. Let two lattices be formal expressed as $LTC1 = < S1, R1 >$ and $LTC2 = < S2, R2 >$. The process of lattice integration needs to take the relationship of nodes in the graphs between both into consideration. And this relationship can be divided into two, the **Equal** relation and the **Dominate** relation.

Definition 1. *Let* $l_1 = < c_1, k_1 >$, $l_2 = < c_2, k_2 >$ *be two security levels.* l_1 *is equal to* l_2, *if and only if* $c_1 = c_2$ *and* $k_1 = k_2$.

Definition 2. *Let* $l_1 = < c_1, k_1 >$, $l_2 = < c_2, k_2 >$ *be two security levels.* l_1 *dominates* l_2, *denoted as* $l_1 \geq l_2$, *if and only if* $c_1 \geq c_2$ *and* $k_1 \supseteq k_2$.

According to these two definitions, during the integration process, if two nodes in Hasse graphs satisfy the equal relation, then we can add two directed lines between the two nodes with each of one direction respectively. And if the nodes in Hasse graph satisfy the dominate relation, then we can add a directed line from the parent node to the child node.

For example, in Fig. 4, assuming that the relationship between $a1$ in *Lattice*1 and $b1$ in *Lattice*2 is equal, we add two lines on the merged Hasse graph, $a1 \rightarrow b1$ and $b1 \rightarrow a1$, as Line 1 and Line 2 in Fig. 5. Assuming that $a2$ in *Lattice*1 and $b1$ in *Lattice*2 satisfy the dominate relation, we add a line from $a2$ to $b1$, $a2 \rightarrow b1$ as Line 3 in Fig. 5.

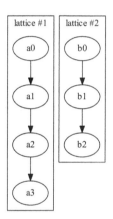

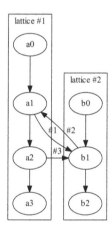

Fig. 4. LTC1 and LTC2   **Fig. 5.** Merging LTC1 and LTC2

The Conflict Management Phase. After adding these lines following the relationship between each of nodes in the Hasse graph, there may exist some redundant lines or conflicting lines. The following step is to remove the lines which are in conflict with the origin lattice graphs.

Firstly, we give some definitions of elements in the Hasse graph:

Definition 3. Path *in a Hasse graph is a sequence of directed lines which connect a sequence of nodes, which are distinct from one another.*

Definition 4. Loop *in a Hasse graph is a special* Path, *of which the start node is the same as the end node, across more than two nodes.*

In addition, we give two relations of the nodes in a Hasse graph, **comparable** and **incomparable**.

Definition 5. *Let s1 and s2 be two nodes in the Hasse graph. s1 and s2 are comparable if and only if there is a path between s1 and s2.*

Definition 6. *Let s1 and s2 be two nodes in the Hasse graph. s1 and s2 are incomparable if and only if there is no a path between s1 and s2.*

Assuming that we have two lattices, Lattice 3 and Lattice 4, they can be represented in Hasse graph, as Fig. 6. And now they need to merge together. If the relationship between a0 and b0 is equal, and a1 dominates b0, b1 dominates a2, following the definition, we can draw an initial merged Hasse graph, as Fig. 7.

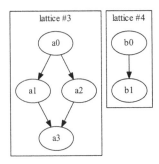

Fig. 6. LTC3 and LTC4

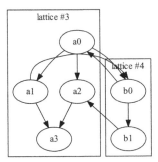

Fig. 7. Merging LTC3 and LTC4

In Fig. 7, it is possible to find out some conflicting issues in the internal graph. The main conflicting problems consist of two cases.

Definition 7. *A path in conflicting is that*

1. *the path is a loop*
2. *the path contains nodes that are incomparable in the original Hasse graph but are comparable in merged Hasse graph.*

More illustrations about the conflict path are as follows.

(1) After adding all the relation lines, the merged graph has a loop. For example, in Fig. 7, there is a loop a0 → a1 → b0 → a0. Under this situation, we can not judge which security level is higher than others. Therefore we need to remove some relation lines.

(2) Another conflict is more implicit. The incomparable nodes in the original Hasse graph become comparable in a merged Hasse graph. For example, in Lattice 3, a1 and a2 cannot be comparable, which means that in Lattice 3, a1 cannot travel to a2 or a2 can not travel to a1. But in the initial merged graph, adding the relation lines, there is a path a1 → b0 → b1 → a2. So, we also need to cut some relation lines to keep the privacy.

To tackle these conflicts, relation lines need to be removed. The removal follows the following rules.

Rule 1: Remove the relation line which appears the most number of times in all the conflict paths.

In a merged Hasse graph, list all the conflict paths and find out the relation line that appears the most number of times, remove it. As illustrated in Fig. 7, there are two conflict paths, $a0 \to a1 \to b0 \to a0$ and $a1 \to b0 \to b1 \to a2$. In these two paths, $a1 \to b0$ appears twice, other relation lines appear once. So the relation line $a1 \to b0$ is to be removed, and get a new merged graph, as Fig. 8. And after the simplification process, merging node $a0$ and $b0$ into one node $b0\_a0$, we can get the final Hasse graph, as Fig. 9.

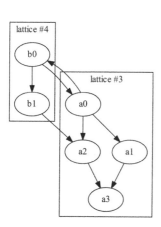

Fig. 8. Rule 1

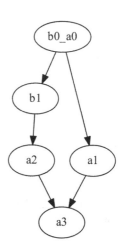

Fig. 9. Simplified Hasse

Rule 2: If all the lines appear the same number of times, remove the line which relates more confidential security levels.

If all relation lines in conflict paths appear the same number of times, the relation line relating to higher security level is to be removed. For example, in Fig. 10, there is only one conflict path $a1 \to b0 \to b1 \to a2$. All the relation lines appear once, but $a1 \to b0$ related to more confidential security level among them, because $b0$ dominates $b1$. Therefore, this relation line is to be removed, resulting in Fig. 11. After simplifying, we can draw the final Hasse graph as Fig. 12.

The Simplification Phase. After deleting some relation lines, there are no conflict lines in a Hasse graph. Lastly, the graph needs to be simplified to remove redundant lines.

Two relations of the paths in the Hasse graph are defined, **equivalent** and **overlapped**.

58   G. Zhao et al.

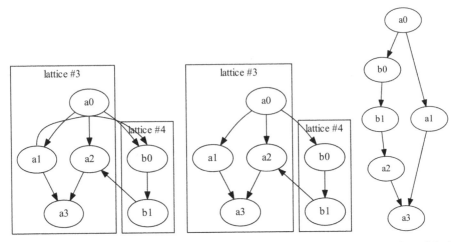

Fig. 10. Incomparable nodes   **Fig. 11.** Rule 2   **Fig. 12.** Simplified Hasse

Definition 8. *Let s1 and s2 be two nodes in a Hasse graph. Paths between s1 and s2 are* equivalent, *if and only if there are two paths directly connecting two nodes, s1 and s2, with each of one direction respectively, as s1 → s2 and s2 → s1.*

Definition 9. *Let s1 and s2 be two nodes in a Hasse graph. Paths between s1 and s2 are* overlapped, *if and only if there are two paths between s1 and s2, one directly connects s1 and s2 and the other one travels through some nodes, as s1 → s2 and s1 → · · · → s2.*

The definition of the redundant lines is as follows:

Definition 10. *Redundant lines are the lines in the paths which are* equivalent *or* overlapped.

The steps of removing the redundant lines are based on these two cases. The rules of simplification are following.

Rule 1: If two paths are equivalent, remove these two paths and merging the nodes connected by these paths into a new node.

If two nodes connect each other, $Level_a → Level_b$ and $Level_b → Level_a$, then these two nodes are merged into a new node, called $Level_{a,b}$. For example, in Fig. 13, there are two nodes a_1 and b_1 which are linked to each other. They can be merged into a new node $a_1\_b_1$, as Fig. 14.

Rule 2: If two paths are overlapped, remove the path which directly connects the start node and the end node.

Let there be a direct path from two nodes, $Level_a → Level_b$. And there is another path travelling across some nodes, $Level_a → Level_1 → Level_2... → Level_3 → Level_b$. According to Rule 2, remove $Level_a → Level_b$. For example,

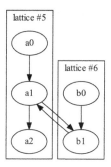

Fig. 13. Equivalence lines

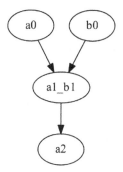

Fig. 14. Rule 1

in Fig. 15, there are two paths in LTC3 and LTC4 from a_0 to b_1, $a_0 \rightarrow b_1$ and $a_0 \rightarrow a_1 \rightarrow b_1$. At this time, $a_0 \rightarrow b_1$ is a redundant line to remove. Then the Hasse graph is clear enough as Fig. 16.

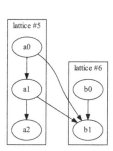

Fig. 15. Overlapped lines

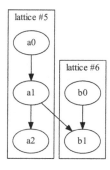

Fig. 16. Rule 2

After removing these redundant lines, it can be proved that the simplified Hasse graph is the same as the redundant Hasse graph.

Theorem 1. *The simplified Hasse graph is the same as the redundant Hasse graph.*

Proof. We would discuss on two cases based on two types of redundant lines.

(1) If two paths connecting $s1$ and $s2$ are equivalent in the redundant graph, after simplification, a merged node s_{new} is generated. The indegree and outdegree of s_{new} is the sum of $s1$ and $s2$. So if the nodes and $s1$ or $s2$ are comparable, those nodes and s_{new} are comparable.

(2) If two paths connecting $s1$ and $s2$ are overlapped in the redundant graph, after simplification, $s1$ and $s2$ are comparable because there is still a path connecting them. So, to sum up, the nodes in the redundant graphs keep the relations among them in the simplified graph.

Algorithm of Merging Two Lattices. Algorithm 1 is to merge two lattices. Function $IsEqual$ is to determine whether the nodes respectively in two lattices are equal. If the answer is yes, then we add two sets of nodes into the relation sets, as to add two lines in the graph. If no, then go to the next steps. Function $IsSub$ is to determine whether one node dominates the other. If it is true, then we add one set (parent node, child node) into the relation set. This is the same as to add a line from parent node to child node. After the initial process, the initial Hasse graph is generated. It needs to be detected whether there are conflicts, like a loop or a path from two incomparable nodes, called Function $Conflict$. According to the conflict removing criteria, some conflict paths will be removed. Finally, the step is to simplify in the light of simplification criteria.

4.3   Mapping Function Translation

Due to the fact that every data source uses BLP to govern the access to the data, a function f_i (i represents the data source i) maps subjects or objects to their security levels. For instance, $f_1(u_1) = \{TS, < k1, k2 >\}$ can be understood that in data source 1, user u_1 has the security level $\{TS, < k1, k2 >\}$.

In policy integration, after integrating lattices, the mapping function from the original security level to a new security level needs to be constructed. The security level can be represented as a node in a Hasse graph. So the nodes in the original Hasse graph can translate to the nodes in the merged Hasse graph. We denote f_i^G, where i is the original LTC_i, as a translation function between each original Lattice and the merged Lattice. f_G^i, as a translation function between the merged Lattice and each original Lattice. The former function is used to translate the original security level into the global security level, while the latter function is used to translate the global security level into the original security level.

Figures 17 and 18 are illustrated two lattices and one merged lattice. Every node in the graph represents a security level. So the nodes in $LTC3$ and $LTC4$ need to be translated to nodes in merged Lattice LTC_G. Table 1 describes a function translation between a lattice and a merged lattice. The first column is the data source number, and the second column is the security level on original Lattice, and the third column is the security level on merged Lattice, and the last column is the function translation between the original lattice and the merged lattice.

Integration process combines the multiple data sets and generates a data base $m_n = Int(\sum_{i=1}^{n} o_i)$. To protect the privacy of each original data set, the security level $f_G(o^G)$ of the data sets o^G in the merged data base m_n is the least upper bound of the security level of data sets. After mapping the original lattice to the merged lattice, let $f_G(o_p^G) = f_p^G(f_p(o_p))$ and $f_G(o_q^G) = f_q^G(f_q(o_q))$ be two global

---

Algorithm 1. Lattice Integration Process

Input: Two Lattice $LTC1$, $LTC2$
Output: Merged Lattice
1: **function** LATTICEINTEG($LTC1$,$LTC2$)
2:      $S \leftarrow S1 \cup S2$
3:      $R \leftarrow R1 \cup R2$
4:      **while** $s_1 \in S1$ **AND** $s_2 \in S2$ **do**
5:          **if** ISEQUAL(s_1,s_2) **then** //Judge whether two nodes are equal
6:              $R' \leftarrow R' + (s_1, s_2) + (s_2, s_1)$
7:          **else**// Judge whether one node dominates the other
8:              **if** ISSUB(s_1,s_2) **then**
9:                  $R' \leftarrow R' + (s_1, s_2)$
10:             **end if**
11:             **if** ISSUB(s_2,s_1) **then**
12:                 $R' \leftarrow R' + (s_2, s_1)$
13:             **end if**
14:         **end if**
15:     **end while**
16:     **while** CONFLICT(R',$R1$,$R2$) **do** // Judge whether there is conflicted
17:         $arc \leftarrow arc + R'$
18:     **end while**
19:     **while** CONFLICT(arc,$R1$,$R2$) **do**
20:         **if** $Count_{Max}(arc) \leq 1$ **then**
21:             $R' \leftarrow R' - arc_{max}$ // Delete the relation line appearing the most number
    of times
22:         **else**
23:             $R' \leftarrow R' -$ HIGHEST(arc) // Delete the line related to the highest secu-
    rity level
24:         **end if**
25:     **end while**
26:     $R' \leftarrow$ SIMPLIFIED(R') // Simplification Process
27:     $R \leftarrow R \cup R'$
28:     **return** $< S, R >$
29: **end function**

Table 1. Mapping function translation

Data source	Security level	Merged security level	Mapping function translation
Lattice 3	a0	b0_a0	$f_3^G(a0) = b0\_a0$, $f_G^3(b0\_a0) = a0$
	a1	a1	$f_3^G(a1) = a1$, $f_G^3(a1) = a1$
	a2	a2	$f_3^G(a2) = a2$, $f_G^3(a2) = a2$
	a3	a3	$f_3^G(a3) = a3$, $f_3^G(a3) = a3$
Lattice 4	b0	b0_a0	$f_4^G(b0) = b0\_a0$, $f_G^4(b0\_a0) = b0$
	b1	b1	$f_4^G(b1) = b1$, $f_G^4(b1) = b1$

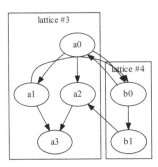

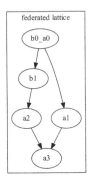

Fig. 17. LTC3 and LTC4   **Fig. 18.** Merging LTC3 and LTC4

security levels which map the lattice p and lattice q to the merged lattice G. If $f_G(o_p^G)$ and $f_G(o_q^G)$ are comparable, which means that there is a path between them. Then the security level of the data sets in the $m_n = Int(o_p, o_q)$ is the higher security level in them $f_G(o^G) = max(f_q^G(f_q(o_q)), f_p^G(f_p(o_p)))$.

In Fig. 18, $b1$ and $a2$ are comparable. When integrating these data sets which label in $a2$ and $b1$, because $b1$ dominates $a2$, the data set in m_n is tagged by $b1$.

But if these two security levels are incomparable, then these nodes in the merged lattice can not be connected. So we need to choose the last common node that can both connect them from top to bottom $f_G(o^G) = min(f_G(o))$, where $f_G(o) \geq f_p^G(f_p(o_p)), f_G(o) \geq f_q^G(f_q(o_q))$. In Fig. 18, $b1$ and $a1$ are incomparable. When integrating the data sets which tag in $a1$ and $b1$ respectively, the last common node connecting them is $b0\_a0$. So the data set in m_n is tagged by $b0\_a0$.

4.4   Matrix Integration

Except for the mandatory access control, we have access control matrix in our model, which is the discretionary access control. The access control matrix integration is to merge the rows and columns with the original matrix. Let $M = SUB \times OBJ$ be the merged matrix and $M1 = subject1 \times object1$ and $M2 = subject2 \times object2$ be two original access matrices. And the column names of the merged matrix is $m_n = Int(object1, object2)$. And the rows index of the merged matrix is $SUB = subject1 \cup subject2$. User can access the integrated data base which merged two data sets, if and only if the user can access both data sets before integration. Therefore, it can be described in a formal way as follows: The value of the merged matrix is $True$, $M_G(sub, obj) = True$, when $sub \in subject1$, $sub \in suject2$ and $obj \in object1$, $obj \in object2$ and $M_1(sub, obj) = True$ and $M_2(sub, obj) = True$. Otherwise, the value is $False$. Algorithm 2 is to merge two access matrixes to generate an integrated access matrix.

Algorithm 2. Matrix Integration

Input: Two Access Matrix$M1$, $M2$
Output: Merged Access Matrix
 1: **function** MATRIXINTEG($M1$,$M2$)
 2:      $SUB \leftarrow subject1 \cup subject2$
 3:      $OBJ \leftarrow m_n = Int(object1, object2)$
 4:      **while** $sub \in SUB$ **do**
 5:          **if** $obj \in obj1$ **and** $obj \in obj2$ **then**
 6:              $M_G(sub, obj) \leftarrow M_1(sub, obj) \wedge M_2(sub, obj)$
 7:          **else**
 8:              $M_G(sub, obj) \leftarrow FALSE$
 9:          **end if**
10:      **end while**
11:      **return** M_G
12: **end function**

---

5   Security Analysis and Proof

The integration of security policies is to ensure that data confidentiality is maintained consistent as that before data are integrated. Hence, the integrated policy needs to make sure that the returned search result does not contain any data from the data sets that are not allowed to access by the specific user.

The security can be formalized as follows. Let O_{RS} be the original data sets of the result sets which return from the integrated data, and let O_{deny} be the unaccessible data sets of the data sources, then $\forall o(o \in O_{RS} \rightarrow o \notin O_{deny})$.

Let m be the integrated data base, generated as $m = Int(\sum_{i=1}^{n} O_i)$, where n is the number of the data sets. The original data sources DS of m is formed by a mapping function, $DS = s\_s(m) = (ds_1, ds_2, \ldots, ds_p)$ and p denotes the number of the data sources. In each data source i, O^i are the total data sets and O^i_{deny} are the unaccessible data sets which are governed by its own policy. The complete confidential data sets are generated $O_{deny} = \bigcup O^i_{deny}$. Let RS be the result set querying on m, generated as $RS = search(u, kwds, m)$, where $kwds$ represent the key words. O_{RS} are the original data sets mapping from RS, generated by $O_{RS} = s\_d(RS)$.

Theorem 2. *For a given user u, u searches $kwds$ on m_n and gets $RS = search(u, kwds, m_n)$. Let O_{RS} be the original data sets of RS and let O_{deny} be the data sets which are not permitted to access by u, then $\forall o(o \in O_{RS} \rightarrow o \notin O_{deny})$*

Proof. We will prove this by induction. The base case is when $n = 2$. That is to say, we would integrate two data sets, $O_n = \{o_1, o_2\}$, generating an integrated data base $m = Int(o_1, o_2)$. According to the mapping function between the data sets and data sources, $DS = s\_s(m) = (ds_1, \ldots, ds_p)$. p, the number of the original data sources, will be discussed on two case: $p = 1$ or $p = 2$.

– If $p = 1$, then the data sets to be merged come from the same data source. For a given user u, if he could access the data sets by searching on the integrated data base, then we have

$$\forall x(x \in O_{RS}) \rightarrow (f_G(u) \geq f_G(x)) \wedge M_G(u, x) = True \qquad (1)$$

where f_G be a mapping function between users/data sets and the merged lattice, and $M_G(u, x)$ be the entry of user u accessing data set x in the access control matrix.

Because o_1 and o_2 come from the same data source ds_1, in the integrated policy $P_G = (f_G, LTC_G, M_G)$, the Lattice and mapping function is the same as those in the policy $P_1 = (f_1, LTC_1, M_1)$ of ds_1, then $LTC_G = LTC_1$ and $f_G = f_1$ and $M_G = M_1$.

And the accessible data sets O_{allow} of ds_1 for user satisfy

$$O_{allow} = \{x | f_1(u) \geq f_1(x) \wedge M_1(u, x) = True\} \qquad (2)$$

According to 1, 2,

$$\forall x(x \in O_{RS}) \rightarrow (f_G(u) \geq f_G(x) \wedge M_G(u, x) = True)$$
$$\Rightarrow (f_1(u) \geq f_1(x) \wedge M_1(u, x) = True)$$
$$\Rightarrow x \in O_{allow}$$

So,

$$O_{RS} \subseteq O_{allow}$$

And because

$$O_{deny} = O - O_{allow},$$

we can get

$$O_{RS} \cap O_{deny} = \emptyset$$

Hence, when $n = 2$ and $p = 1$, we prove that the theorem is true.

- If $p = 2$, then the data sets to be merged come from two data sources. Based on our model, for a given user u, if he could access the data sets by searching on the integrated data base, then we have:

$$\forall x(x \in O_{RS}) \rightarrow (f_G(u) \geq f_G(x) \wedge M_G(u, x) = True) \qquad (3)$$

where f_G is a mapping function on the merged Lattice between users or data sets and security levels, and $M_G(u, x)$ is the value of the access matrix.
If $f_G(o_1)$ and $f_G(o_2)$ are comparable, then

$$f_G(x) = max(f_G(o_1), f_G(o_2)) \qquad (4)$$

Or if $f_G(o_1)$ and $f_G(o_2)$ are incomparable, then

$$f_G(x) \geq min(f_G(o)) \qquad (5)$$

where $f_G(o) \geq f_G(o_1)$ and $f_G(o) \geq f_G(o_2)$
In addition, we can get

$$M_G(u, x) = True \Rightarrow M_1(u, x) = True \wedge M_2(u, x) = True \qquad (6)$$

And the accessible data sets O^1_{allow} of ds_1 for user satisfy

$$O^1_{allow} = \{x | f_1(u) \geq f_1(x) \wedge (u, x) = True\} \tag{7}$$

And the accessible data sets O^2_{allow} of ds_1 for user satisfy

$$O^2_{allow} = \{x | f_2(u) \geq f_2(x) \wedge (u, x) = True\} \tag{8}$$

As we have a translation mapping function f^i_G, which translates between f_G and f_i, so we can get

$$\forall i \in O_{RS}, f_i(y) = f^i_G(f_G(y)) \tag{9}$$

where y are users or data sets in the system.

According to 3, 4, 5, 6, 7, 9, we can get

$$\forall x (x \in O_{RS} \wedge x \in O^1) \rightarrow f_G(u) \geq f_G(x) \wedge M_G(u, x) = True$$
$$\Rightarrow f^1_G(f_G(u)) \geq f^1_G(f_G(x)) \wedge M_1(u, x) = True$$
$$\Rightarrow f_1(u) \geq f_1(x) \wedge M_1(u, x) = True$$
$$\Rightarrow x \in O^1_{allow}$$

In a similar way, according to 3, 4, 5, 6, 8, 9, we can get

$$\forall x (x \in O_{RS} \wedge x \in O^2) \rightarrow x \in O^2_{allow}$$

Therefore, we can get

$$O_{RS} \subseteq (O^1_{allow} \cup O^2_{allow})$$

Because O^1 and O^2 are independent data sets, we can get

$$O_{deny} = O^1_{deny} \cup O^2_{deny}$$
$$= (O^1 - O^1_{allow}) \cup (O^2 - O^2_{allow})$$
$$= (O^1 \cup O^2) - (O^1_{allow} \cup (O^2_{allow}))$$

So,

$$O_{RS} \cap O_{deny} = \emptyset$$

Hence, when $n = 2$ and $p = 1$, we prove that the theorem is true.

Assuming that when $n = k - 1$, the theorem holds, we now consider the case when $n = k$. When $n = k - 1$, we denotes that $DS_{p'} = s\_s(m_{k-1}) = \{ds_1, ds_2, \ldots, ds_{p'}\}$ and the number of original data source is $p = p'$. As known, the process of integrating more than two data sets together,

$$m_k = Int(\sum_{i=1}^{k} o_i) = Int(m_{k-1}, o_k)$$

where

$$m_{k-1} = Int(\sum_{i=1}^{k-1} o_i)$$

From integration process definition, when $n = k - 1$, a integrated data base $m_{k-1} = Int(\sum_{i=1}^{k-1} o_i)$. This integrated data is assumed to keep confidentiality. When it merges with the kth data set, we need to consider whether the original data source of kth data set is in DS_{k-1}. So this issue would be discussed on two cases: $p = p'$ and $p = p' + 1$.

- $p = p'$. It means that the original data source of o_k is in DS_{k-1}. Then there is no new integrated Lattice, and it can be translated to the condition of $m = 2$ and $p = 1$. So it can be proved that the theorem is true on the basis of the condition of $m = 2$ and $p = 1$.
- $p = p' + 1$. It means that the original data source of o_k is not in DS_{k-1}. Then a new integrated Lattice would be generated as our model. It is similar to the condition of $m = 2$ and $p = 2$. So it can be proved that the theorem is true on the basis of the condition of $m = 2$ and $p = 2$.

To sum up, we have shown that the theorem holds for $n = 2$, and we have also shown that if the theorem holds for $n = k - 1$ then it holds for $n = k$. We can therefore state that it holds for all $n(n \geq 2)$.

6   Conclusions

6.1   Work Summary

This work investigates the issues of protecting integrated data in the scenario of searching over multiple data sources. For data sources protected by BLP security policies, this work proposes a model of protecting the integrated data. The proposed model generates a new BLP security policy according to the integration of data from the original data sources. The new BLP security policy is generated by merging the original BLP security policies' security level lattices as well as constructing a new access control matrix consistent with the original ones. The merging of security level lattices also tackles the issue of security conflicts, where the merging is based on Hasse graphs. Theoretical security analysis and proof are conducted, whose result shows that the proposed model is capable of generating appropriate BLP security policy to protect integrated data from multiple data sources, retaining the confidentiality of these data as required by the security policies of the data sources.

6.2   Contributions

The contributions of this work are as follows.

Firstly, we identify the need of protecting the confidentiality of integrated data from multiple data sources, especially in the scenario of searching over the integrated data.

Secondly, we formalize the process of data integration, to articulate the data flow when they are integrated from multiple data sources and demonstrate how data can be access via integration.

Thirdly, we propose a method for integrating BLP security policies in accordance with the integration of data. The proposed method combines security levels from multiple BLP security policies, tackles the conflicts of the security levels, and constructs a new access control matrix. New mapping functions for mapping users to security clearance and for mapping data to security levels are also constructed. This proposed method can then generate a new BLP security policy for protecting the integrated data.

Fourthly, we conduct a formal security analysis and proof to show that the proposed model is capable of generating the corresponding new BLP security policy, and the new BLP security policy is capable of protecting the data confidentiality of integrated data according to the data's original security policies.

6.3   Future Work

This work is not without limitations. Currently, this work assumes that data sources are protected based on BLP security policies. While in reality, there are also many data sources protected based on RBAC security policies. To address this limitation, future work will explore possible ways to work with other security models, such as RBAC model.

Acknowledgements. Research is supported in part by the China MOE-China Mobile Research Fund (No. MCM20121051), China MOE Doctoral Research Fund (No. 201344 07120017), Guangdong Nature Science Fund (No. S2012030006242), Guangdong Modern Information Service Fund (GDEID2012IS063).

References

1. Agrawal, D., Das, S., El Abbadi, A.: Big data and cloud computing: current state and future opportunities. In: Proceedings of the 14th International Conference on Extending Database Technology, pp. 530–533. ACM (2011)
2. Alodib, M.: An approach to automating the integration of the access control policies for web services. In: 2013 14th ACIS International Conference on Software Engineering, Artificial Intelligence, Networking and Parallel/Distributed Computing (SNPD), pp. 181–187. IEEE (2013)
3. Ardagna, C.A., De Capitani di Vimercati, S., Paraboschi, S., Pedrini, E., Samarati, P., Verdicchio, M.: Expressive and deployable access control in open web service applications. IEEE Trans. Serv. Comput. **4**(2), 96–109 (2011)
4. Cao, N., Wang, C., Li, M., Ren, K., Lou, W.: Privacy-preserving multi-keyword ranked search over encrypted cloud data. IEEE Trans. Parallel Distrib. Syst. **25**(1), 222–233 (2014)
5. Cruz, I.F., Gjomemo, R., Orsini, M.: A secure mediator for integrating multiple level access control policies. In: Lovrek, I., Howlett, R.J., Jain, L.C. (eds.) KES 2008, Part II. LNCS (LNAI), vol. 5178, pp. 354–362. Springer, Heidelberg (2008)

6. Famaey, J., De Turck, F.: Federated management of the future internet: status and challenges. Int. J. Netw. Manag. **22**(6), 508–528 (2012)
7. Feng, D.G., Zhang, M., Zhang, Y., Xu, Z.: Study on cloud computing security. J. Softw. **22**(1), 71–83 (2011)
8. Google: Gsa notes from the field: Security. http://static.googleusercontent.com/media/www.google.com/en/us/support/enterprise/static/gsa/docs/deployment/en/GSASecurity.pdf. Accessed Jan 2015
9. Haddad, M., Hacid, M., Laurini, R.: Data integration in presence of authorization policies. In: 2012 IEEE 11th International Conference on Trust, Security and Privacy in Computing and Communications (TrustCom), pp. 92–99. IEEE (2012)
10. Halevy, A., Ives, Z.: Principles of Data Integration. Elsevier, Amsterdam (2012)
11. Hu, Y.J., Yang, J.J.: A semantic privacy-preserving model for data sharing and integration. In: Proceedings of the International Conference on Web Intelligence, Mining and Semantics, pp. 9:1–9:12. ACM (2011)
12. Li, M., Yu, S., Cao, N., Lou, W.: Authorized private keyword search over encrypted data in cloud computing. In: 2011 31st International Conference on Distributed Computing Systems (ICDCS), pp. 383–392. IEEE (2011)
13. Marchant, R.L.: Common access control terminology used in multilevel security systems. In: Proceedings of the Information Systems Educators Conference (2012). ISSN: 2167–1435
14. Pan, L., Xu, Q.: Visualization analysis of multi-domain access control policy integration based on tree-maps and semantic substrates. Intell. Inf. Manag. **4**(5), 188–193 (2012)
15. Rao, P., Lin, D., Bertino, E., Li, N., Lobo, J.: An algebra for fine-grained integration of xacml policies. In: Proceedings of the 14th ACM Symposium on Access Control Models and Technologies, pp. 63–72. ACM (2009)
16. Rao, P., Lin, D., Bertino, E., Li, N., Lobo, J.: Fine-grained integration of access control policies. Comput. Secur. **30**(2–3), 91–107 (2011)
17. Ren, K., Wang, C., Wang, Q., et al.: Security challenges for the public cloud. IEEE Internet Comput. **16**(1), 69–73 (2012)
18. Sellami, M., Gammoudi, M.M., Hacid, M.S.: Secure data integration: a formal concept analysis based approach. In: Decker, H., Lhotská, L., Link, S., Spies, M., Wagner, R.R. (eds.) DEXA 2014, Part II. LNCS, vol. 8645, pp. 326–333. Springer, Heidelberg (2014)
19. Su, M., Li, F., Shi, G., Li, L.: An action based access control model for multi-level security. Int. J. Secur. Appl. (IJSIA) **6**(2), 359–366 (2012)
20. Tankard, C.: Big data security. Netw. Secur. **2012**(7), 5–8 (2012)
21. Thorleuchter, D., Van den Poel, D.: Improved multilevel security with latent semantic indexing. Expert Syst. Appl. **39**(18), 13462–13471 (2012)
22. Wang, C., Cao, N., Li, J., Ren, K., Lou, W.: Secure ranked keyword search over encrypted cloud data. In: 2010 IEEE 30th International Conference on Distributed Computing Systems (ICDCS), pp. 253–262. IEEE (2010)
23. Watson, P., Little, M.: Multi-level security for deploying distributed applications on clouds, devices and things. In: 2014 IEEE 6th International Conference on Cloud Computing Technology and Science (CloudCom), pp. 380–385. IEEE (2014)
24. Watson, P.: A multi-level security model for partitioning workflows over federated clouds. J. Cloud Comput. **1**(1), 1–15 (2012)
25. Wikipedia: Hasse diagram. http://en.wikipedia.org/wiki/Hasse_diagram. Accessed Jan 2015
26. Xue, H., Zhang, Y., Guo, Z.: A multilevel security model for private cloud. Chin. J. Electron. **23**(2), 232–235 (2014)

Security Analysis of Two Identity Based Proxy Re-encryption Schemes in Multi-user Networks

Xu An Wang[1,3](✉), Jianfeng Ma[2], Xiaoyuan Yang[3], and Yuechuan Wei[3]

[1] School of Telecommunications Engineering,
Xidian University, Xi'an, People's Republic of China
wangxazjd@163.com
[2] School of Cyber Engineering, Xidian University, Xi'an, People's Republic of China
[3] Engineering University of Chinese Armed Police Force,
Xi'an, People's Republic of China

Abstract. In proxy re-encryption (PRE), a semi-trusted proxy can convert a ciphertext originally intended for Alice into one which can be decrypted by Bob, while the proxy can not know the underlying plaintext. In multi-use PRE schemes, the ciphertext can be transformed from Alice to Bob and to Charlie and so on. Due to its ciphertext transformation property, it is difficult to achieve chosen ciphertext security for PRE, especially for multi-use PRE. IBE is a new kind of public-key encryption where the recipient's public key is an arbitrary string that represents the recipient's identity. Identity based proxy re-encryption (IBPRE) is a primitive combing the feature of IBE and PRE. In 2010 Wang *et al.* has proposed a multi-use unidirectional CCA-secure identity based proxy re-encryption (IBPRE) scheme, and in 2011 Luo *et al.* has proposed an unidirectional identity based proxy re-encryption scheme. Unfortunately, we show these two proposals are not secure and thus can not be applied directly in multi-user networks.

Keywords: Identity based proxy re-encryption · Insecurity · Security model · Transitivity property · Attack

1 Introduction

Blaze et al. (1998) [1] introduced the concept of proxy re-encryption (PRE) in 1998. The goal of proxy re-encryption is to securely enable the re-encryption of ciphertexts from one key to another, without relying on trusted parties. According to the direction of transformation, PRE schemes can be classified into bidirectional schemes and unidirectional schemes. Also according to the times the transformation can apply on the ciphertext, PRE schemes can be classified into single-hop schemes and multi-hop schemes. PRE was recently investigated by Ateniese et al. (2005, 2006) [2,3]. They proposed the first construction of unidirectional PRE and demonstrated several applications of PRE. However, their schemes can only achieve chosen plaintext security. It is important to

© Springer-Verlag Berlin Heidelberg 2015
N.T. Nguyen et al. (Eds.): Transactions on CCI XIX, LNCS 9380, pp. 69–88, 2015.
DOI: 10.1007/978-3-662-49017-4_5

achieve chosen ciphertext security (CCA-security) for many practical applications. In CCS'07, Canetti and Hohenberger (2007) [4] proposed the first CCA-secure bidirectional PRE. However, their schemes suffer from collusion attacks. In PKC'08, Libert and Vergnaud (2008) [5] proposed the first unidirectional PRE scheme which is replayable chosen ciphertext attack (RCCA) secure and collusion resistant in the standard model.

Identity Based Proxy Re-encryption. In ACNS'07, Green and Ateniese (2007) [6] proposed the first IBPRE schemes. They defined the algorithms and security models for IBPRE, and constructed their scheme by using a variant of the efficient Dodis/Ivan key splitting approach to the settings with a bilinear map [7]. But it was found later that it cannot resist collusion attack [8]. This scheme is also the first multi-use unidirectional IBPRE (PRE) scheme. However their scheme can not achieve CCA-security. In ISC'07, Chu and Tzeng (2007) [9] claimed to propose the first CCA-secure IBPRE scheme in the standard model based on Waters' IBE. However, this scheme is not efficient due to the structure of Waters' IBE and Green's paradigm. In Pairing'07, Matsuo (2007) [10] proposed proxy re-encryption schemes for identity based system. They constructed an IBPRE scheme. But recently it was shown that this scheme has some flaws [11]. In Inscrypt'08, [12] proposed the inter-domain IBPRE scheme. They concern on constructing PRE between different domains in identity based setting. In SDM'08, Tang (2008) [13] constructed a type-and-identity based PRE and discussed its application in health care. Lai et al. [14] gave new constructions on IBPRE with master secret security based on identity-based mediated encryption.

Recently, as a new computing paradigm, cloud computation received great attention from the academy society and the industry society. Many research work has been done on how to increase the efficiency and strength the security of cloud computation [17, 20, 21], among which the (identity based) proxy re-encryption or attribute based encryption techniques are very promising one [15, 16, 18, 19, 22]. In Information Science, Wang et al. [23] claimed to propose the first multi-use CCA-secure unidirectional IBPRE scheme, but we shall show that this scheme is yet not CCA-secure either. And at ICISC'11, Luo et al. [24] proposed a unidirectional IBPRE scheme, but we give some analysis to show that their scheme may have some shortcomings, and thus can not be applied in multi-user networks.

1.1   Our Contribution

We give the formal security models for multi-use CCA-secure IBPRE, then we give concrete attacks on a IBPRE scheme proposed by Wang et al. [23] in this security model. Thus their scheme is not CCA-secure. By exploiting some features of Luo et al. (2011)'s IBPRE scheme [24], we argue their scheme's security is not as secure as they claimed.

1.2   Organization

We organize our paper as follows. In Sect. 2.1, we first recall the definition and security models of multi-use CCA-secure IBPRE, and then review of an IBPRE scheme proposed by Wang *et al.* Then we give several attacks on this scheme in Sect. 3. In Sect. 4, we review of a single-hop IBPRE scheme proposed by Luo *et al.* and its security model. and then we give some security analysis on their scheme in Sect. 5. In the last Sect. 6, we give our conclusion.

2 Review of a Multi-use CCA-secure IBPRE Scheme

2.1 Definition and Security Model

Definition 1. *A multi-use and unidirectional identity-based proxy re-encryption scheme is a tuple of algorithms (*Setup, Extract, RKExtract, Encrypt, Reencrypt, Decrypt*) as follows:*

1. Setup(1^k). On input a security parameter k, the algorithm outputs the system's public parameters ($params$) which are distributed to users, and the master secret key (msk) which is kept private to the PKG. The system parameters include a description of a finite message space $\mathcal{M}$ and a description of a ciphertext space $\mathcal{C}$.
2. Extract($params, msk, id$). On input an identity $id \in \{0,1\}$ and the master secret key (msk), the algorithm outputs a decryption key sk_{id} corresponding to the user with identity id.
3. RKExtract($params, sk_{id_i}, id_j$). On input a secret key sk_{id_i} and identity $id_j \in \{0,1\}$, the algorithm outputs a unidirectional re-encryption key from id_i to id_j as $rk_{id_i \to id_j}$.
4. Encrypt($params, id, m$). On input a set of public parameters, an identity $id \in \{0,1\}^*$, and a plaintext $m \in \mathcal{M}$, the algorithm outputs the first-level ciphertext $c_{id}^{(1)}$, the encryption of m under identity id.
5. ReEncrypt($params, rk_{id_i \to id_j}, c_{id_i}^{(l)}$). On input an lth-level ciphertext $c_{id_i}^{(l)}$ under identity id, and a re-encryption key $rk_{id_i \to id_j}$, the algorithm outputs an $(l+1)$th-level re-encrypted ciphertext $c_{id_j}^{(l+1)}$, where $c_{id_j}^{(l+1)}$ is under identity id_j.
6. Decrypt($params, sk_{id}, c_{id}^{(l)}$). ($l \geq 1$). The algorithm decrypts the lth-level ciphertext $c_{id}^{(l)}$ using the secret key sk_{id}, and outputs $m \in \mathcal{M}$ or $\perp$.

We say that a unidirectional and multi-use IBPRE scheme is consistent if for any valid identities id_i and id_j, ($i \neq j$) their secret keys sk_{id_i}, sk_{id_j}, generated by Extract, the corresponding re-encryption key $rk_{id_i \to id_j}$, generated by RKExtract, and an lth-level ($l \geq 1$) ciphertext $c_{id_i}^{(l)}$ under identity id_i output by Encrypt or Reencrypt, the following equations hold for $\forall m \in \mathcal{M}$:

$$\mathsf{Decrypt}(params, sk_{id_i}, \mathsf{Encrypt}(params, id_i, m)) = m;$$

$$\mathsf{Decrypt}(params, sk_{id_j}, \mathsf{ReEncrypt}(params, rk_{id_i \to id_j}, c_{id_i}^{(l)})) = m, l \geq 1$$

Definition 2 (CCA-Security). *A multi-use unidirectional IBPRE scheme is CCA-secure if the advantage of any PPT adversary $\mathcal{A}$ in the following game played between a challenger $\mathcal{C}$ and $\mathcal{A}$ is negligible in the security parameter k. Note that we work in the static corruption model, where the adversary should decide the corrupted users before the game starts.*

Let E be an IBPRE scheme defined as above. We consider the following game, in which a p.p.t. adversary $\mathcal{A}$ is involved, denoted by $Game_{E,A}^{IND-ID-CCA2}$:

- **Setup Phase:** Challenger runs Setup to get $(params, msk)$, and gives $params$ to $\mathcal{A}$. It keeps the master secret key to itself.
- **Find Phase:** $\mathcal{A}$ makes queries $q_1, \cdots, q_k$, where q_i, $i = 1, \cdots, k$ is one of:
  1. On $\mathcal{A}$'s any query of the form $(extract, id)$, the challenger responds by running algorithm Extract to generate the corresponding private key sk_{id}. The challenger then returns sk_{id} to $\mathcal{A}$.
  2. On $\mathcal{A}$'s any query of the form $(rkextract, id_i, id_j)$, where $id_i \neq id_j$, the challenger returns $rk_{id_i \rightarrow id_j} =$ RKExtract$(params,$ Extract$(params, msk, id_i), id_j)$ to $\mathcal{A}$.
  3. On $\mathcal{A}$'s any query of the form $(reencrypt, id_i, id_j, c_{id_i}^{(l)}, l \geq 1$, the challenge first derives a re-encryption key $rk_{id_i \rightarrow id_j} =$ RKExtract$(params,$ Extract$(params, msk, id_i), id_j)$, then returns $c_{id_j}^{(l+1)}=$ReEncrypt$(params,$ $rk_{id_i \rightarrow id_j}, c_{id_i}^{(l)}$ to $\mathcal{A}$.
  4. On $\mathcal{A}$'s any query of the form $(decrypt, id_i, c_{id_i}^{(l)})(l \geq 1)$, the challenger returns $m=$Decrypt$(params,$ Extract$(params, msk, id_i), c_{id_i}^{(l)})$ to $\mathcal{A}$.

  These queries maybe asked adaptively, i.e., each query may depend on the replies to $q_1, \cdots, q_{i-1}$. At the end of this phase, $\mathcal{A}$ presents his select $id^* \in \{0, 1\}$ and two same length messages $(m_0, m_1) \in \mathcal{M}^2$. Note that $\mathcal{A}$ is not permitted to choose id^* such that trivial decryption is possible using keys extracted during this phase (i.e., by using extracted re-encryption keys to translate from id^* to some identity for which $\mathcal{A}$ holds a decryption key).
- **Challenge Phase:** Once $\mathcal{A}$ decides that the Find phase is over, he presents his choice (id^*, m_0, m_1). The challenger randomly choose $b \leftarrow_R \{0, 1\}$ and computes $c^* =$ Encrypt$(params, id^*, m_b)$, and gives c^* to $\mathcal{A}$ as the challenge ciphertext.
- **Guess Phase:** $\mathcal{A}$ continues to make queries $q_{k+1}, \cdots, q_n$ as in the Find phase, with the following restrictions. Let $\mathcal{C}$ be a set of ciphertext/identity pairs, initially containing the single pair (c^*, id^*). For all $(c, *) \in \mathcal{C}$, and for all rk given to $\mathcal{A}$ or can be computed by $\mathcal{A}$, let C' be the set of all possible values derived via (one or more) consecutive calls to Reencrypt:
  1. $\mathcal{A}$ is not permitted to launch any query of the form $(decrypt, id, c)$, where $(c, id) \in C \cup C'$.
  2. $\mathcal{A}$ is not permitted to launch any query of the form $(extract, id)$ or $(rkextract, id_i, id_j)$ that would permit trivial decryption of any ciphertext in $C \cup C'$.
  3. $\mathcal{A}$ is not permitted to launch any query of the form $(reencrypt, id_i, id_j, c)$, where $\mathcal{A}$ possesses the keys to trivially decrypt ciphertexts under id_j and $(c, id_i) \in C \cup C'$. On successful execution of any re-encryption query, let c' be the result and add the pair (c', id_j) to the set C.

  These queries maybe asked adaptively as in the Find Phase. At the end of this phase, $\mathcal{A}$ outputs his guess b', where $b = b'$, then $\mathcal{A}$ wins the game.

We define adversary $\mathcal{A}$'s advantage in attacking PRE as

$$Adv_{PRE}^{IND-ID-CCA2}(k) = |Pr[b = b'] - \frac{1}{2}|.$$

If for all probabilistic polynomial time algorithms $\mathcal{A}$, $Adv_{E,A}^{IND-ID-CCA2}(k)$ is negligible with respect to k, then we say that an IBPRE scheme is secure against adaptive chosen ciphertext and identity attack.

2.2 The Wang et al.'s IBPRE Scheme

Here we review a recently IBPRE scheme proposed by Wang et al. (2010) [23] in Information Science. Let 1^k be the security parameter and $(q, g, \mathbb{G}_1, \mathbb{G}_T, e) \leftarrow BSetup(1^k)$, and $Sig = (G, S, V)$ be a strongly unforgeable signature scheme, where $l = l(k)$ denotes the length of the verification keys output by $\mathbb{G}(1^k)$. Moreover, we assume that any given key in Sig has a negligible chance of being sampled.

1. **Setup**(1^k): Let the message space be $\mathcal{M} = \{0,1\}^{k_0}$ such that $k_0 < k$ and both $\frac{1}{2^{k_0}}$ and $\frac{1}{2^{k-k_0}}$ are negligible. Let $\mathbb{G}_1$ and $\mathbb{G}_2$ be two multiplicative groups of the same prime order q with $\lceil log_2^q \rceil = k$, such that the discrete logarithm problems in both $\mathbb{G}_1$ and $\mathbb{G}_2$ are intractable. Suppose that g is a generator of $\mathbb{G}_1$ and $e : \mathbb{G}_1 \times \mathbb{G}_1 \rightarrow \mathbb{G}_T$ is a bilinear pairing map. Let $H_1 : \{0,1\}^* \rightarrow \mathbb{G}_1$, $H_2 : \{0,1\}^* \rightarrow \mathbb{G}_1$, $H_3 : \mathbb{G}_T \rightarrow \mathbb{G}_1$, $H_4 : \{0,1\}^* \rightarrow \mathbb{G}_1$ be four hash functions. To generate the scheme parameters, select master secret key $msk = s$ randomly from Z_p^*, and output the system parameters as

$$param = (\mathcal{M}, k, k_0, \mathbb{G}_1, \mathbb{G}_T, q, e, H_1, H_2, H_3, H_4, g, g^s)$$

2. **Extract** $(params, msk, id)$. To generate a decryption key for identity $id \in \{0,1\}^*$, compute $sk_{id} = H_1(id)^s$ and send it to the user with the identity id via a secure channel.

3. **RKExtract**$(params, sk_{id_i}, id_j)$. To generate a re-encryption key for id_i's proxy P_i, id_i selects r_i, X_i from Z_q^* and $\mathbb{G}_T$ randomly first, then computes

$$R_1^{(i)} = g^{r_i}, R_2^{(i)} = X_i \cdot e(g^s, H_1(id_j))^{r_i}, R_3^{(i)} = svk_{P_i},$$
$$R_4^{(i)} = H_2(svk_{P_i})^{r_i}, R_5^{(i)} = H_2(R_1^{(i)}||R_2^{(i)}||R_3^{(i)})^{r_i},$$
$$R_6^{(i)} = H_3(X_i) \cdot sk_{id_i}^{-1}$$

where svk_{P_i} is a publicly available verification key of P_i. Finally, id_i outputs $rk_{id_i \rightarrow id_j} = (R_1^{(i)}, R_2^{(i)}, R_3^{(i)}, R_4^{(i)}, R_5^{(i)}, R_6^{(i)})$

4. **Encrypt**$(params, id, m)$. To encrypt a message $m \in \mathcal{M}$ under identity id, select r from Z_q^* and $\sigma \in \{0,1\}^{k-k_0}$ randomly, then compute

$$c_{1,1} = g^r, c_{1,2} = (m||\sigma) \cdot e(g^s, H_1(id)^r),$$
$$c_{1,3} = H_2(m||\sigma||c_{1,1}) \cdot g^{r\sigma},$$
$$c_{1,4} = U = H_4(c_{1,1}||c_{1,2}||c_{1,3})^r$$

and output the ciphertext as $c_{id}^{(1)} = (c_{1,1}, c_{1,2}, c_{1,3}, c_{1,4})$.

5. **ReEncrypt**$(params, rk_{id_i \to id_j}, c^{(l)}_{id_i})$.
   - To re-encrypt a first-level ciphertext $c^{(l)}_{id}$, denoted by $c^{(l)}_{id}$, do the following
     (a) Parse $c^{(l)}_{id}$ as $(c_{1,1}, c_{1,2}, c_{1,3}, c_{1,4})$ and $rk_{id_i \to id_j}$ as $(R^{(i)}_1, R^{(i)}_2, R^{(i)}_3, R^{(i)}_4, R^{(i)}_5, R^{(i)}_6)$.
     (b) Check if $e(g, c_{1,4}) = e(c_{1,1}, H_4(c_{1,1}, c_{1,2}, c_{1,3}))$. If not, return $\perp$.
     (c) Otherwise, compute $C = (c'_{1,1}, c'_{1,2}, c'_{1,3}, c'_{1,4}, c'_{2,1}, c'_{2,2}, c'_{2,3}, c'_{2,4}, c'_{2,5})$, where $c'_{1,1} = c_{1,1}, c'_{1,2} = c_{1,2} \cdot e(c_{1,1}, R^{(1)}_6), c'_{1,3} = c_{1,3}, c'_{1,4} = c_{1,4}, c'_{2,1} = R^{(1)}_1, c'_{2,2} = R^{(1)}_2, c'_{2,3} = R^{(1)}_3, c'_{2,4} = R^{(1)}_4, c'_{2,5} = R^{(1)}_5$.
     (d) Suppose ssk_{P_i} is the signature key of id_i's proxy P_i corresponding to $R^{(1)}_3$.
     (e) Run the signing algorithm $S(ssk_{P_i}, (c'_{1,1}, c'_{1,2}, c'_{1,3}, c'_{1,4}, c'_{2,1}, c'_{2,3}, c'_{2,4}, c'_{2,5}))$ to generate a signature on the ciphertext tuple $(c'_{1,1}, c'_{1,2}, c'_{1,3}, c'_{1,4}, c'_{2,1}, c'_{2,3}, c'_{2,4}, c'_{2,5})$, and denote the signature as S_1.
     (f) Output the ciphertext $c^{(2)}_{id_j} = (C, S_1)$.
   - To re-encrypt an lth-level ($l \geq 1$) ciphertext $c^{(l)}_{id_i}$, denoted by $c^{(1)}_{id_i}$, do:
     (a) Parse $c^{(l)}_{id_l}$ as $(c_{1,1}, \cdots, c_{l,1}, c_{l.2}, c_{l,3}, c_{l,4}, c_{l,5}, S_{l-1})$, and $rk_{id_i \to id_j}$ as $(R^{(l)}_1, R^{(l)}_2, R^{(l)}_3, R^{(l)}_4, R^{(l)}_5, R^{(l)}_6)$
     (b) Check if $e(g, c_{l,5}) = e(c_{l,1}, H_2(c_{l,1} \| c_{l,2} \| c_{l,3}))$, if not, return $\perp$.
     (c) For $\forall k \in [2, l]$, check whether

$$e(g, c_{k,4}) = e(c_{k,1}, H_2(c_{k,3}))$$

and

$$V(svk_{P_{k-1}}, S_{k-1}, (c_{1,1}, \cdots, c_{k,1}, c_{k,2}, c_{k,3}, c_{k,4}, c_{k,5})) = 1$$

whenever one of them fails, return $\perp$. Otherwise, do the next:
     (d) Compute $C = (c'_{1,1}, \cdots, c'_{l,1}, c'_{l,2}, c'_{l,3}, c'_{l,4}, c'_{l,5}, c'_{l,6}, c'_{l+1,1}, c'_{l+1,2}, c'_{l+1,3}, c'_{l+1,4}, c'_{l+1,5})$, where $c'_{l,2} = c_{l,2} \cdot e(c_{l,1}, R^{(l)}_6), c'_{l+1,1} = R^{(l)}_1, c'_{l+1,2} = R^{(l)}_2, c'_{l+1,3} = R^{(l)}_3, c'_{l+1,4} = R^{(l)}_4, c'_{l+1,5} = R^{(l)}_5$, and all other elements remain unchanged.
     (e) Suppose ssk_{P_i} is the signature key of id_i's proxy P_i corresponding to $R^{(l)}_3$.
     (f) Run the signing algorithm $S(ssk_{P_i}, c'_{1,1}, \cdots, c'_{l+1,1}, c'_{l+1,2}, c'_{l+1,3}, c'_{l+1,4}, c'_{l+1,5})$ to generate a signature on the ciphertext tuple $(c'_{1,1}, \cdots, c'_{l+1,1}, c'_{l+1,2}, c'_{l+1,3}, c'_{l+1,4}, c'_{l+1,5})$, and denote the signature as S_l.
     (g) Output the ciphertext $c^{(l+1)}_{id_j} = (C, S_l)$.

6. **Decrypt**$(params, sk_{id}, c_{id}^{(l)})$. If $c_{id}^{(l)}$ can not be parsed as $(c_{1,1}, c_{1,2}, c_{1,3}, c_{1,4})$ for a first-level ciphertext, or $(c_{1,1}, \cdots, c_{1,4}, \cdots, c_{l,1}, \cdots, c_{l,6})$ for an l-level ciphertext $(l \geq 1)$, then return $\perp$; otherwise continue the following process:

(a) If $l = 1$, then perform

    i. Verify that $e(g, c_{1,4}) = e(c_{1,1}, H_2(c_{1,1}||c_{1,2}||c_{1,3}))$. If not, return $\perp$.

    ii. Otherwise, compute $m' = \frac{c_{1,2}}{e(c_{1,1}, sk_{id})}$.

    iii. Parse m' as (m, σ).

    iv. Verify that $c_{1,3} = H_2(m||\sigma||c_{1,1}) \cdot c_{1,1}^\sigma$. If not, return $\perp$, otherwise, output m.

(b) Otherwise, if $l \geq 1$, perform

    i. Check if $e(g, c_{l,5}) = e(c_{l,1}, H_2(c_{l,1}||c_{l,2}||c_{l,3}))$, if not, return $\perp$.

    ii. For $\forall k \in [2, l]$, check

        A. whether $e(g, c_{k,4}) = e(c_{k,1}, H_2(c_{k,3}))$. If not, return $\perp$.

        B. $V(svk_{P_{k-1}}, S_{k-1}, (c_{1,1}, \cdots, c_{k,1}, c_{k,3}, c_{k,4}, c_{k,5})) = 1$

    whenever one of them fails, output $\perp$, otherwise, do the next.

    iii. Compute $X_{l-1} = \frac{c_{l,2}}{e(c_{l,1}, sk_{id})}$.

    iv. For i from $l - 2$ down to 1, compute $X_i = \frac{c_{i+1,2}}{e(c_{i+1,1}, H_3(X_{i+1}))}$

    v. Compute $m' = \frac{c_{1,2}}{e(c_{1,1}, H_3(X_1))}$.

    vi. Parse m' as (m, σ).

    vii. If $c_{1,3} \neq H_2(m||\sigma||c_{1,1}) \cdot c_{1,1}^\sigma$, return $\perp$; otherwise, return m.

In this IBPRE scheme, the signing key $sskp_{P_i}$ of id_i's proxy is not explicitly included in the the re-encryption key $rk_{id_i \to id_j}$. However, $sskp_{P_i}$ should indeed be implicitly included in $rk_{id_i \to id_j}$. Otherwise, without $sskp_{P_i}$, algorithm **ReEncrypt** cannot be properly performed.

3 Our Attack

The authors claimed that their multi-use IBPRE scheme is CCA-secure. However, in this section, we show this is not true. Concretely, there exists a polynomial time adversary $\mathcal{A}$ who has non-negligible advantage against the CCA security of this multi-use IBPRE scheme.

3.1 Attack I

We can see the intuition of the attack I in Fig. 1.

Adversary $\mathcal{A}$ works as follows:

1. In **Setup phase**, adversary $\mathcal{A}$ obtains the system parameters $params$.
2. In **Find phase**, $\mathcal{A}$ needs not issue any query.
3. In **Challenge phase**, $\mathcal{A}$ outputs a target identity id^* and two equal-length plaintexts m_0, m_1. Then $\mathcal{A}$ is given a challenge ciphertext $c^* = $ **Encrypt** $(params, id^*, m_d)$. According to **Encrypt** algorithm, $c^* = (c_{1,1}^*, c_{1,2}^*, c_{1,3}^*, c_{1,4}^*)$ should be of the following forms:

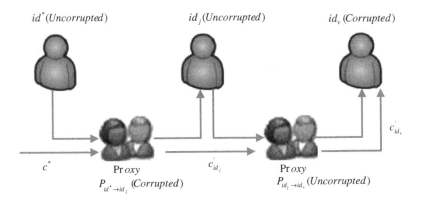

Fig. 1. Attack I

$$c_{1,1}^* = g^{r^*}, c_{1,2}^* = (m_d||\sigma^*) \cdot e(g^s, H(id^*)^{r^*}),$$
$$c_{1,3}^* = H_2(m_d||\sigma^*||c_{1,1}^*) \cdot g^{r^*\sigma^*},$$
$$c_{1,4}^* = H_4(c_{1,1}^*||c_{1,2}^*||c_{1,3}^*)^{r^*}.$$

4. In **Guess phase**, $\mathcal{A}$ corrupts the proxy P_{i^*} who can re-encrypts ciphertexts from id^* to another identity id_j, and $\mathcal{A}$ obtains $(ssk_{P_{i^*}}, R_1^{(1)}, R_2^{(1)}, R_3^{(1)}, R_4^{(1)}, R_5^{(1)}, R_6^{(1)})$, where

$$R_1^{(1)} = g^{r_{i^*}}, R_2^{(1)} = X_{i^*} \cdot e(g^s, H_1(id_j))^{r_{i^*}},$$
$$R_3^{(1)} = svk_{P_{i^*}}, R_4^{(1)} = H_2(svk_{P_{i^*}})^{r_{i^*}},$$
$$R_5^{(1)} = H_2(R_1^{(1)}||R_2^{(1)}||R_3^{(1)})^{r_{i^*}},$$
$$R_6^{(1)} = H_3(X_{i^*}) \cdot sk_{id^*}^{-1}.$$

Note that, it is legal for $\mathcal{A}$ to issue this query, since $\mathcal{A}$ does not corrupt user id_j, and thus it does not permit trivial decryption of any ciphertext in $(\mathcal{C} \cup \mathcal{C}')$. Next, $\mathcal{A}$ picks a number $\theta \in Z_q^*$, and generates an (ill-formed) 2-level ciphertext $c'_{id_j} = (c'_{1,1}, \bar{c}'_{1,2}, c'_{1,3}, c'_{1,4}, c'_{2,1}, c'_{2,2}, c'_{2,3}, c'_{2,4}, c'_{2,5}, S_1)$, where

$$c'_{1,1} = c_{1,1}^* = g^{r^*},$$
$$\bar{c}'_{1,2} = \theta \cdot c_{1,2}^* \cdot e(c_{1,1}^*, R_6^{(1)}) = \theta \cdot (m_d||\sigma^*) \cdot e(g^s, H_3(X_{i^*})),$$
$$c'_{1,3} = c_{1,3}^* = H_2(m_d||\sigma^*||c_{1,1}^*) \cdot g^{r^*\sigma^*},$$
$$c'_{1,4} = c_{1,4}^* = H_4(c_{1,1}^*||c_{1,2}^*||c_{1,3}^*)^{r^*},$$
$$c'_{2,1} = R_1^{(1)} = g^{r_{i^*}},$$
$$c'_{2,2} = R_2^{(1)} = X_{i^*} \cdot e(g^s, H_1(id_j))^{r_{i^*}},$$
$$c'_{2,3} = R_3^{(1)} = svk_{P_{i^*}},$$

$$c'_{2,4} = R_4^{(1)} = H_2(svk_{P_{i*}})^{r_{i*}},$$
$$c'_{2,5} = R_5^{(1)} = H_2(R_1^{(1)}||R_2^{(1)}||R_3^{(1)})^{r_{i*}},$$
$$S_1 = \mathcal{S}(ssk_{P_{i*}}, (c'_{1,1}, \bar{c}'_{1,2}, c'_{1,3}, c'_{1,4}, c'_{2,1}, c'_{2,3}, c'_{2,4}, c'_{2,5})).$$

Note that, adversary $\mathcal{A}$ generates $\bar{c}'_{1,2} = \theta \cdot c^*_{1,2} \cdot e(c^*_{1,1}, R_6^{(1)})$ instead of $\bar{c}'_{1,2} = c^*_{1,2} \cdot e(c^*_{1,1}, R_6^{(1)})$. Obviously, c'_{id_j} is an *ill-formed* ciphertext.

5. $\mathcal{A}$ corrupts another identity id_ν to obtain sk_{id_ν}, and issues a re-encryption query on c'_{id_j} from identity id_j to identity id_ν. Note that it is legal for $\mathcal{A}$ to issue these queries, since c'_{id_j} is an ill-formed ciphertext, and these query does not permit trivial decryption of any ciphertext in $(\mathcal{C} \cup \mathcal{C}')$. Note also that, although c'_{id_j} is an ill-formed ciphertext, it still can passe the validity checks in algorithm **ReEncrypt**, since the following equations hold:

$$e(g, c'_{2,5}) = e(c'_{2,1}, H_2(c'_{2,1}||c'_{2,2}||c'_{2,3})),$$
$$e(g, c'_{2,4}) = e(c'_{2,1}, H_2(c'_{2,3})),$$
$$\mathcal{V}(svk_{P_{i*}}, S_1, (c'_{1,1}, \bar{c}'_{1,2}, c'_{1,3}, c'_{1,4}, c'_{2,1}, c'_{2,3}, c'_{2,4}, c'_{2,5})) = 1.$$

So, the re-encryption oracle has to return the re-encrypted ciphertext c'_{id_ν} of c'_{id_j} to $\mathcal{A}$. According to the re-encryption algorithm, $c'_{id_\nu} = (c'_{1,1}, \bar{c}'_{1,2}, c'_{1,3}, c'_{1,4}, c'_{2,1}, \bar{c}'_{2,2}, c'_{2,3}, c'_{2,4}, c'_{2,5}, c'_{3,1}, c'_{3,2}, c'_{3,3}, c'_{3,4}, c'_{3,5}, S_2)$ should be of the following forms:

$$\bar{c}'_{2,2} = c'_{2,2} \cdot e(c'_{2,1}, R_6^{(2)}), c'_{3,1} = R_1^{(2)}, c'_{3,2} = R_2^{(2)},$$
$$c'_{3,3} = R_3^{(2)}, c'_{3,4} = R_4^{(2)}, c'_{3,5} = R_5^{(2)},$$
$$S_2 = \mathcal{S}(ssk_{P_j}, c'_{1,1}, \bar{c}'_{1,2}, c'_{1,3}, c'_{1,4}, c'_{2,1}, \bar{c}'_{2,2}, c'_{2,3}, c'_{2,4},$$
$$c'_{2,5}, c'_{3,1}, c'_{3,2}, c'_{3,3}, c'_{3,4}, c'_{3,5}).$$

where $R_1^{(2)} = g^{r_j}, R_2^{(2)} = X_j \cdot e(g^s, H_1(id_\nu))^{r_j}, R_3^{(2)} = svk_{P_j}, R_4^{(2)} = H_2(svk_{P_j})^{r_j}, R_5^{(2)} = H_2(R_1^{(2)}||R_2^{(2)}||R_3^{(2)})^{r_j}, R_6^{(2)} = H_3(X_j) \cdot sk_{id_j}^{-1}$. So we have

$$c'_{1,1} = g^{r^*}, \bar{c}'_{1,2} = \theta \cdot (m_d||\sigma^*) \cdot e(g^s, H_3(X_{i*})),$$
$$c'_{1,3} = H_2(m_d||\sigma^*||c^*_{1,1}) \cdot g^{r^*\sigma^*},$$
$$c'_{1,4} = H_4(c^*_{1,1}||c^*_{1,2}||c^*_{1,3})^{r^*}, c'_{2,1} = g^{r_{i*}},$$
$$\bar{c}'_{2,2} = X_{i*} \cdot e(g^s, H_3(X_j)), c'_{2,3} = svk_{P_{i*}},$$
$$c'_{2,4} = H_2(svk_{P_{i*}})^{r_{i*}}, c'_{2,5} = H_2(R_1^{(1)}||R_2^{(1)}||R_3^{(1)})^{r_{i*}},$$
$$c'_{3,1} = g^{r_j}, c'_{3,2} = X_j \cdot e(g^s, H_1(id_\nu))^{r_j}, c'_{3,3} = svk_{P_j},$$
$$c'_{3,4} = H_2(svk_{P_j})^{r_j}, c'_{3,5} = H_2(R_1^{(2)}||R_2^{(2)}||R_3^{(2)})^{r_j},$$
$$S_2 = \mathcal{S}(ssk_{P_j}, c'_{1,1}, \bar{c}'_{1,2}, c'_{1,3}, c'_{1,4}, c'_{2,1}, \bar{c}'_{2,2}, c'_{2,3}, c'_{2,4},$$
$$c'_{2,5}, c'_{3,1}, c'_{3,2}, c'_{3,3}, c'_{3,4}, c'_{3,5}).$$

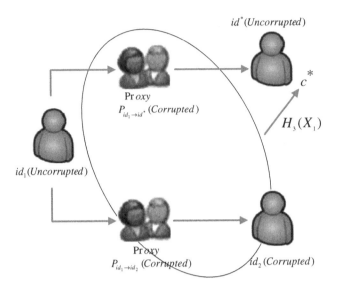

Fig. 2. Attack II

6. Now, with $c'_{id_\nu} = (c'_{1,1}, \bar{c}'_{1,2}, c'_{1,3}, c'_{1,4}, c'_{2,1}, \bar{c}'_{2,2}, c'_{2,3}, c'_{2,4}, c'_{2,5}, c'_{3,1}, c'_{3,2}, c'_{3,3}, c'_{3,4},$ $c'_{3,5}, S_2)$ and using sk_{id_ν}, adversary $\mathcal{A}$ can recovers the underlying plaintext m_d as follows:

$$x_j = \frac{c'_{3,2}}{e(c'_{3,1}, sk_{id_\nu})} = \frac{X_j \cdot e(g^s, H_1(id_\nu))^{r_j}}{e(g^{r_j}, H_1(id_\nu)^s)},$$

$$X_{i^*} = \bar{c}'_{2,2} \cdot e(g^s, H_3(X_j))^{-1}$$
$$= X_{i^*} \cdot e(g^s, H_3(X_j)) \cdot e(g^s, H_3(X_j))^{-1},$$

$$(m_d||\sigma^*) = \frac{\bar{c}'_{1,2} \cdot e(g^s, H_3(X_i^*))^{-1}}{\theta}$$
$$= \frac{\theta \cdot (m_d||\sigma^*) \cdot e(g^s, H_3(X_{i^*})) \cdot e(g^s, H_3(X_{i^*}))^{-1}}{\theta}.$$

Knowing m_d, adversary $\mathcal{A}$ can certainly break this IBPRE scheme.

3.2   Attack II

We can see the intuition of the attack II in Fig. 2.

Adversary $\mathcal{A}$ works as follows:

1. In **Setup Phase**, adversary $\mathcal{A}$ obtains the public parameters *params* from the challenger $\mathcal{C}$.
2. In **Find Phase**, adversary $\mathcal{A}$ needs not issue any queries.

3. In **Challenge Phase**, adversary $\mathcal{A}$ returns a challenged identity id^*, and two equal-length plaintexts m_0, m_1. Then challenger $\mathcal{C}$ picks $b \in_R \{0, 1\}$, sets the challenge ciphertext to $c^* = \mathsf{ReEncrypt}(params, id^*, \mathsf{Encrypt}(params, id_1, m_b))$, and gives c^* to $\mathcal{A}$. Recall that $\mathcal{A}$'s goal is to correctly guess the value b. Denote $c^* = (c_{1,1}^*, c_{1,2}^*, c_{1,3}^*, c_{2,1}^*, c_{2,2}^*, c_{2,3}^*, c_{2,4}^*, c_{2,5}^*)$ where

$$c_{1,1}^* = g^r, c_{1,2}^* = (m_b || \sigma) \cdot e(g^r, H_3(X_1)),$$
$$c_{1,3}^* = H_2(m_b || \sigma || c_{1,1}^*) \cdot g^{r\sigma},$$
$$c_{2,1}^* = g^{r_1}, c_{2,2}^* = R_2 = X_1 \cdot e(g^s, H_1(id^*))^{r_1},$$
$$c_{2,3}^* = svk_P, c_{2,4}^* = H_2(svk_P)^{r_1},$$
$$c_{2,5}^* = H_2(R_1 || R_2 || R_3)^{r_1}$$

Here we assume the proxy between id_1 and id^* is P, $c = \mathsf{Encrypt}(params, id_1, m_b) = (c_{1,1}, c_{1,2}, c_{1,3}, c_{1,4})$, and $rk_{id_1 \to id^*} = (R_1, R_2, R_3, R_4, R_5, R_6)$ where

$$c_{1,1} = g^r, c_{1,2} = (m_b || \sigma) \cdot e(g^s, H_1(id_1)^r),$$
$$c_{1,3} = H_2(m_b || \sigma || c_{1,1}) \cdot g^{r\sigma},$$
$$c_{1,4} = H_4(c_{1,1} || c_{1,2} || c_{1,3})^r$$
$$R_1 = g^{r_1}, R_2 = X_1 \cdot e(g^s, H_1(id^*))^{r_1},$$
$$R_3 = svk_P, R_4 = H_2(svk_P)^{r_1},$$
$$R_5 = H_2(R_1 || R_2 || R_3)^{r_1}, R_6 = H_3(X_1) \cdot sk_{id_1}^{-1}$$

4. In **Guess Phase**, adversary $\mathcal{A}$ does as the follows:
   (a) First he corrupts the proxy P between id_1 and id^*, and he will get

   $$R_1 = g^{r_1}, R_2 = X_1 \cdot e(g^s, H_1(id^*))^{r_1},$$
   $$R_3 = svk_P, R_4 = H_2(svk_P)^{r_1},$$
   $$R_5 = H_2(R_1 || R_2 || R_3)^{r_1}, R_6 = H_3(X_1) \cdot sk_{id_1}^{-1}$$

   (b) Then he issues a query of the form $(rekeygen, id_1, id_2)$, and he will get $rk_{id_1 \to id_2} = (R_1', R_2', R_3', R_4', R_5', R_6')$ where

   $$R_1' = g^{r_2}, R_2' = X_1' \cdot e(g^s, H_1(id_2))^{r_2},$$
   $$R_3' = svk_{P'}, R_4' = H_2(svk_{P'})^{r_2},$$
   $$R_5' = H_2(R_1' || R_2' || R_3')^{r_2}, R_6' = H_3(X_1') \cdot sk_{id_1}^{-1}$$

   Here we assume the proxy between id_1 and id_2 is P'.
   (c) Then he corrupts P' and id_2 and he will get sk_{id_1} as follows

   $$X_1' = \frac{R_2'}{e(R_1', sk_{id_2})} = \frac{X_1' \cdot e(g^s, H_1(id_2))^{r_2}}{e(g^{r_2}, H_1(id_2)^s)}$$
   $$sk_{id_1} = \frac{H_3(X_1')}{R_6'} = \frac{H_3(X_1')}{H_3(X_1') \cdot sk_{id_1}^{-1}}$$

(d) Now knowing sk_{id_1} and R_6, he can get $H_3(X_1) = R_6 \cdot sk_{id_1}$, then he computes $m' = \frac{c^*_{1,2}}{e(c^*_{1,1}, H_3(X_1))}$, parse m' as (m_b, σ) and check $c^*_{1,3} = H_2(m_b||\sigma||c^*_{1,1}) \cdot (c^*_{1,1})^\sigma$ holding, thus he will get m_b.

Now, knowing the plaintext m_b, adversary $\mathcal{A}$ can certainly know the underlying bit b chosen by the challenger in **Challenge Phase**, and hence always wins the game.

Note here we just attack the single-hop variant of their scheme, it can be easily extended to the multi-hop variants. The adversary $\mathcal{A}$ actually is a corrupted proxy P between id_1 and id^* colluding with another proxy P' between id_1 and id_2 where id_2 is also corrupted. Note the adversary $\mathcal{A}$ does not know the first level ciphertext c corresponding to c^* for it does not know $H_4(c^*_{1,1}||c^*_{1,2}||c^*_{1,3})^r$, and thus $(c^*_{1,1}, c^*_{1,2}, c^*_{1,3})$ is not in the set $C \cup C'$ defined in the above security model and this is not a trivial attack.

3.3 On Collusion Resistance

Collusion resistant has widely been accepted as a fundamental security requirement for unidirectional PRE scheme. [23] did not discuss this requirement. Here, we indicate that their scheme fails to satisfy this requirement. Concretely, if the proxy P_i and the delegatee id_j collude, they can easily derive the delegator id_i's secret key as follows:

$$X_i = \frac{R_2^{(i)}}{e(R_1^{(i)}, sk_{id_j})} = \frac{X_i \cdot e(g^s, H_1(id_j))^{r_i}}{e(g^{r_i}, H_1(id_j)^s)} = \frac{X_i \cdot e(g^s, H_1(id_j))^{r_i}}{e(g^s, H_1(id_j))r_i},$$

$$sk_{id_i} = \frac{H_3(X_i)}{R_6^{(i)}} = \frac{H_3(X_i)}{H_3(X_i) \cdot sk_{id_i}^{-1}}.$$

4 Review of Another IBPRE Scheme

4.1 Definition of Single-Hop Unidirectional IBPRE

The definition for single-hop unidirectional IBPRE is almost the same as the definition of the multi-hop unidirectional IBPRE except that, there is no requirement on the re-encryption property of the re-encrypted ciphertext. The security model is more easy than the multi-hop IBPRE for we do not consider the complex case of corruption between proxies and the delegatees.

Definition 3. *A single-hop unidirectional IBPRE scheme consists of the following algorithms:* Setup, KeyGen, ReKeyGen, Enc1, Enc2, ReEnc, Dec1, Dec2.

1. Setup(1^λ). This algorithm takes the security parameter as input, outputs a public key PK, a master secret key MK.
2. KeyGen(MK, I). This algorithm takes MK and an identity I as input, outputs a secret key SK_I associated with I.

3. ReKeyGen(SK_I, I'). This algorithm takes a secret key SK_I and an identity I' as input, outputs a re-encryption key $RK_{I \to I'}$.
4. Enc2(PK, M, I). This algorithm takes PK, a message M, and an identity I as input, outputs a second level ciphertext CT_I.
5. Enc1(PK, M, I'). This algorithm takes PK, a message M, and an identity I' as input, outputs a first level ciphertext $CT_{I'}$.
6. ReEnc($CT_I, RK_{I \to I'}$). This algorithm takes a ciphertext CT_I encrypted to I and a re-encryption key $RK_{I \to I'}$ as input, outputs a ciphertext $CT_{I'}$ encrypted to I'.
7. Dec2(CT_I, SK_I, I). This algorithm takes a second level ciphertext CT_I and SK_I associated with I as input, outputs the message M or the error symbol $\perp$ if CT_I is invalid.
8. Dec1($CT_{I'}, SK_{I'}, I'$). This algorithm takes a first level ciphertext $CT_{I'}$ and $SK_{I'}$ associated with I' as input, outputs the message M or the error symbol $\perp$ if $CT_{I'}$ is invalid.

Correctness. A single-hop unidirectional IBPRE should satisfy the following requirements:

$$Dec(Enc1(PK, M, I'), SK_{I'}) = M,$$

$$Dec(Enc2(PK, M, I), SK_I) = M$$

$$Dec(ReEnc(Enc(PK, M, I), RK_{I \to I'}, SK_{I'})) = M.$$

4.2 Security Model for Single-Hop Unidirectional IBPRE

In fact, there are nine properties that a "good" proxy re-encryption scheme should satisfy [2,3]. But until now, almost no PRE schemes simultaneously satisfy these properties. Here we just consider the security model for the first level ciphertext and the second level ciphertext.

Definition 4. *We define the security of a single-hop unidirectional IBPRE scheme at the second level according to the following* IND-2PrID-ATK *game, where* $ATK \in \{CPA, CCA\}$.

1. Setup. Run the Setup algorithm and give PK to the adversary $\mathcal{A}$.
2. Phase 1. $\mathcal{A}$ makes the following queries:
   - Extract(I): $\mathcal{A}$ submits an identity I for a KeyGen query, return the corresponding secret key SK_I.
   - RKExtract(I, I'): $\mathcal{A}$ submits an identity pair (I, I') for a ReKeyGen query, return the re-encryption key $RK_{I \to I'}$.
   If $ATK = CCA$, $\mathcal{A}$ can make the additional queries:
   - Reencrypt(CT_I, I, I'): $\mathcal{A}$ submits a second level ciphertext CT_I encrypted for I and an identity I' for a ReEnc query, the challenger gives the adversary the re-encrypted ciphertext $CT_{I'} = ReEnc(CT_I, RK_{I \to I'}))$ where $RK_{I \to I'} = ReKeyGen(SK_I, I')$ and $SK_I = KeyGen(MK, I)$.

- Decrypt(CT_I, I): $\mathcal{A}$ submits a first level ciphertext CT_I encrypted for I for a Dec_1 query, return the corresponding plaintext $M = Dec1(CT_I, SK_I)$, where $SK_I = KeyGen(MK, I)$.
3. **Challenge.** $\mathcal{A}$ submits a challenge identity I and two equal length messages M_0, M_1 to $\mathcal{B}$, If the queries
   - Extract(I^*);
   - RKExtract(I^*, I') and Extract(I') for any identity I'.
   are never made, then flip a random coin b and give the ciphertext $CT^* = Enc_2(PK, M_b, I^*)$ to $\mathcal{A}$.
4. **Phase 2.** Phase 1 is repeated with the restriction that $\mathcal{A}$ cannot make the following queries:
   (a) Extract(I^*);
   (b) RKExtract(I^*, I') and Extract(I') for any identity I';
   (c) Reencrypt(CT^*, I^*, I') and Extract(I') for any identity I';
   (d) Decrypt($CT_{I'}, I'$) for any identity I', where $CT_{I'} = ReEnc(CT^*, I^*, I')$.
5. **Guess.** $\mathcal{A}$ outputs its guess b' of b.

The advantage of $\mathcal{A}$ in this game is defined as $Adv_{\mathcal{A}} = |Pr[b' = b] - 1/2|$ where the probability is taken over the random bits used by the challenger and the adversary. If no probabilistic polynomial time adversary $\mathcal{A}$ has a non-negligible advantage in winning the IND-2PrID-ATK game, we say that a single-hop unidirectional IBPRE scheme is IND-2PrID-ATK secure, where $ATK \in \{CPA, CCA\}$.

Definition 5. *The security of a single-hop unidirectional IBPRE scheme at the first level is defined according to the following $IND - 1PrID - ATK$ game, where $ATK \in \{CPA, CCA\}$*

1. **Setup.** Run the Setup algorithm and give PK to the adversary $\mathcal{A}$.
2. **Phase 1.** $\mathcal{A}$ makes the following queries:
   - Extract (I): $\mathcal{A}$ submits an identity I for a KeyGen query, return the corresponding secret key SK_I.
   - RKExtract(I, I'): $\mathcal{A}$ submits an identity pair (I, I') for a ReKeyGen query, return the re-encryption key $RK_{I \to I'}$.
   If $ATK = CCA$, $\mathcal{A}$ can make the additional queries:
   - Decrypt (CT_I, I): $\mathcal{A}$ submits a first level ciphertext CT_I encrypted to I for a Dec_1 query, return the corresponding plaintext $M = Dec_1(CT_I, SK_I)$, where $SK_I = KeyGen(MK, I)$.
3. **Challenge.** $\mathcal{A}$ submits a challenge identity I^* and two equal length messages M_0, M_1 to $\mathcal{B}$. If the query Extract(I^*) is never made, then $\mathcal{C}$ flips a random coin b and passes the ciphertext $CT^* = Enc_1(PK, M_b, I^*)$ to $\mathcal{A}$.
4. **Phase 2.** Phase 1 is repeated with the restriction that $\mathcal{A}$ cannot make the following queries:
   - Extract(I^*).
   - Decrypt(CT^*, I^*).
5. **Guess.** $\mathcal{A}$ outputs its guess b' of b.

The advantage of $\mathcal{A}$ in this game is defined as $Adv_{\mathcal{A}} = |Pr[b' = b] - \frac{1}{2}|$ where the probability is taken over the random bits used by the challenger and the adversary. We say that a single-hop unidirectional IBPRE scheme is IND-1PrID-ATK secure, where $ATK \in \{CPA, CCA\}$, if no probabilistic polynomial time adversary $\mathcal{A}$ has a non-negligible advantage in winning the IND-1PrID-ATK game.

4.3   Luo et al.'s IBPRE Scheme

1. **Setup**(1^λ): Given the security parameter, this algorithm first gets a bilinear group G of order $N = p_1 p_2 p_3$ where p_1 and p_2 are distinct primes. Let $\mathbb{G}_{p_i}$ denote the subgroup of order p_i in $\mathbb{G}$. It then chooses $a, b, c, d, \alpha, \beta, \gamma \in \mathbb{Z}_N$ and $g \in \mathbb{G}_{p_1}$ randomly. Next it computes $u_1 = g^a, h_1 = g^b, u_2 = g^c, h_2 = g^d, w = g^\beta$ and $v = g^\gamma$. The public parameters are published as $PK = (N, g, u_1, h_1, u_2, h_2, w, v, e(g,g)^\alpha)$, The master secret key MK is $\{\alpha, \beta, \gamma, a, b, c, d\}$ and a generator of $\mathbb{G}_{p_3}$. The identity space is Z_N and the message space is $\mathbb{G}_T$.

2. **KeyGen**(MK, I): Given an identity $I \in \mathbb{Z}_N$, this algorithm chooses $r, t, t', x, y, z \in \mathbb{Z}_N$ and $R_3, R_3', \hat{R}_3, \hat{R}_3' \in \mathbb{G}_{p_3}$ randomly, and computes $D_1 = g^\alpha (u_1^I h)^r R_3, D_2 = g^r R_3', E_1 = \frac{c+x}{aI+b}, E_2 = g^{\beta x}, F_1 = \frac{d+y}{aI+b}, F_2 = g^{\beta y}, Z_1 = \frac{z}{aI+b}, Z_2 = g^{\beta z}, K_1 = \frac{t}{\beta(cI+d)}, K_2 = g^\alpha g^{t+\gamma t'} \hat{R}_3, K_3 = g^{t'} \hat{R}_3'$. We require that the PKG always use the same random value t for I. This can be accomplished by using a pseudo-random function (PRF) or an internal log to ensure consistency. The secret key is $SK_I = (D_1, D_2, E_1, E_2, F_1, F_2, Z_1, Z_2, K_1, K_2, K_3)$.

3. **ReKeyGen**(SK_I, I'): Given a secret key $SK_I = (D_1, D_2, E_1, E_2, F_1, F_2, Z_1, Z_2, K_1, K_2, K_3)$ for I and an identity $I' \le I$, this algorithm chooses $k_1, k_2 \in \mathbb{Z}_N$ randomly and computes $rk_1 = (E_1 + k_1 Z_1)I' + (F_1 + k_2 Z_1), rk_2 = (E_2 \cdot Z_2^{k_1})^{I'}(F_2 Z_2^{k_2})$. The re-encryption key is $RK_{I \to I'} = (rk_1, rk_2)$.

4. **Enc**$_2$(PK, M, I): To encrypt a message $M \in G_T$ for an identity I, this algorithm chooses $s \in Z_N$ randomly and computes $C = Me(g,g)^{\alpha s}, C_1 = (u_1^I h_1)^s, C_2 = g^s, C_3 = v^s$. The second level ciphertext is $CT_I = (C, C_1, C_2, C_3)$.

5. **Enc**$_1$(PK, M, I): To encrypt a message $M \in \mathbb{G}_T$ for an identity I, this algorithm chooses $s \in Z_N$ randomly and computes $C = Me(g,g)^{\alpha s}, C_1' = e(u_2^I h_2, w)^s, C_2 = g^s, C_3 = v^s$. The first level ciphertext is $CT_I = (C, C_1', C_2, C_3)$.

6. **ReEnc**($CT_I, RK_{I \to I'}$): Given a second level ciphertext $CT_I = (C, C_1, C_2, C_3)$ and a re-encryption key $RK_{I \to I'} = (rk_1, rk_2)$, this algorithm computes $C_1' = e(C_1, w)^{rk_1} e(C_2, rk_2)^{-1}$. The re-encrypted ciphertext is $CT_I' = (C, C_1', C_2, C_3)$.

7. **Dec**$_2$(CT_I, SK_I): Let $CT_I = (C, C_1, C_2, C_3)$ be a second level ciphertext for identity I, it can be decrypted as $M = C \frac{e(D_2, C_1)}{e(D_1, C_2)}$.

8. **Dec**$_1$(CT_I, SK_I): Let $CT_I = (C, C_1', C_2, C_3)$ be a first level ciphertext for identity I, it can be decrypted as $M = C(C_1')^{K_1} \frac{e(K_3, C_3)}{e(K_2, C_2)}$.

Correctness at Second Level

$$\frac{e(K_3, C_3)}{e(K_2, C_2)} = \frac{e(g^r R_3', (u_1^I h_1)^s)}{e(g^\alpha ((u_1^I h_1)^r) R_3, g^s)} = e(g,g)^{-\alpha s}$$

Correctness at First Level

$$(C_1')^{K_1} \frac{e(K_3, C_3)}{e(K_2, C_2)} = e(u_2^I h_2, w)^{\frac{st}{\beta(cI+d)}} \frac{e(g^{t'} \hat{R}_3', g^{\gamma s})}{e(g^\alpha g^{t+\gamma t'} \hat{R}_3', g^s)} = e(g,g)^{-\alpha s}$$

5   Security Analysis

5.1   Attack by Using the Insecure ReKeyGen Algorithm

We show the adversary can generate any re-encryption key $RK_{I \to I*}$ for arbitrary identity I^* by just querying ReKeyGen oralce three (even can be reduced to two) times, the attacker works as following:

1. First the challenger makes the Extract query on the identity I, then the Key-Gen algorithm generates the private keys for identity I as $D_1 = g^{\alpha}(u_1^I h)^r R_3$, $D_2 = g^r R_3', E_1 = \frac{c+x}{aI+b}, E_2 = g^{\beta x}, F_1 = \frac{d+y}{aI+b}, F_2 = g^{\beta y}, Z_1 = \frac{z}{aI+b}, Z_2 = g^{\beta z}, K_1 = \frac{t}{\beta(cI+d)}, K_2 = g^{\alpha}g^{t+\gamma t'} \hat{R}_3, K_3 = g^{t'} \hat{R}_3'$.
2. Next the adversary makes the ReKeyGen queries on the identities I_1 and I_2, then he will get the following re-encryption keys:

$$rk_1 = (E_1 + k_1 Z_1)I_1 + (F_1 + k_2 Z_1), rk_2 = (E_2 \cdot Z_2^{k_1})^{I_1}(F_2 Z_2^{k_2}) \quad (1)$$

$$rk_1' = (E_1 + k_1' Z_1)I_2 + (F_1 + k_2' Z_1), rk_2' = (E_2 \cdot Z_2^{k_1'})^{I_2}(F_2 Z_2^{k_2'}) \quad (2)$$

Then the adversary can get

$$rk_1 - rk_1' = E_1(I_1 - I_2) + Z_1 \left((k_1 I_1 - k_1' I_2) + (k_2 - k_2')\right) \quad (3)$$

$$rk_2 - rk_2' = E_2^{(I_1 - I_2)} Z_2^{\left((k_1 I_1 - k_1' I_2) + (k_2 - k_2')\right)} \quad (4)$$

3. The adversary continues on making the ReKeyGen query on another identity I_3, then he will get

$$rk_1'' = (E_1 + k_1'' Z_1)I_3 + (F_1 + k_2'' Z_1), \quad (5)$$

$$rk_2'' = (E_2 \cdot Z_2^{k_1''})^{I_3}(F_2 Z_2^{k_2''}) \quad (6)$$

4. Assume the adversary want to forge a re-encryption key $RK_{I \to I*}$ for I^*, he then do the following:

$$\Delta = (3) \cdot \frac{I^* - I_3}{I_1 - I_2} \quad (7)$$

$$= E_1(I^* - I_3) + \frac{Z_1 \left((k_1 I_1 - k_1' I_2) + (k_2 - k_2')\right)(I^* - I_3)}{I_1 - I_2} \quad (8)$$

$$\Omega = (4)^{\frac{I^* - I_3}{I_1 - I_2}} = E_2^{(I^* - I_3)} Z_2^{\left((k_1 I_1 - k_1' I_2) + (k_2 - k_2')\right) \cdot \frac{I^* - I_3}{I_1 - I_2}} \quad (9)$$

5. The adversary can derive a forged (but correct) re-encryption key $RK_{I \to I*}$ as following

$$rk_1^* = \Delta + (5) \quad (10)$$

$$= E_1 I^* + F_1 + \quad (11)$$

$$Z_1 \left[((k_1 I_1 - k_1' I_2) + (k_2 - k_2')) \frac{I^* - I_3}{I_1 - I_2} + k_1'' I_3 + k_2'' \right], \quad (12)$$

$$rk_2^* = \Omega \cdot (6) = E_2^{I^*} F_2 Z_2^{\left[((k_1 I_1 - k_1' I_2) + (k_2 - k_2')) \frac{I^* - I_3}{I_1 - I_2} + k_1'' I_3 + k_2''\right]} \quad (13)$$

let $R_1 = \left[((k_1 I_1 - k_1' I_2) + (k_2 - k_2')) \frac{I^* - I_3}{I_1 - I_2} + k_1'' I_3 + k_2'' \right]$, then

$$T_1 = Z_1 \left[((k_1 I_1 - k_1' I_2) + (k_2 - k_2')) \frac{I^* - I_3}{I_1 - I_2} + k_1'' I_3 + k_2'' \right] \qquad (14)$$

$$= Z_1 R_1 \qquad (15)$$

$$T_2 = Z_2^{\left[((k_1 I_1 - k_1' I_2) + (k_2 - k_2')) \frac{I^* - I_3}{I_1 - I_2} + k_1'' I_3 + k_2'' \right]} = Z_2^{R_1} \qquad (16)$$

then we can verify that

$$e(C_1^{T_1}, w) = e((u_1^I h_1)^{s Z_1 R_1}, g^\beta) = e(g^{s z R_1}, g^\beta) \qquad (17)$$

$$e(C_2, T_2) = e(g^s, g^{\beta z R_1}) \qquad (18)$$

$$e(C_1^{T_1}, w) = e(C_2, T_2) \qquad (19)$$

$$e(C_1^{E_1 I^* + F_1}, w) = e(C_1^{E_1 I^* + F_1}, g^\beta) \qquad (20)$$

$$= e((u_1^{I^*} h_1)^s, w) e(E_2^{I^*} F_2, g^s) \qquad (21)$$

Thus $RK_{I \to I^*}$ is a correct re-encryption key.

If we let I_3 be I_1 or I_2, then the adversary can generate any re-encryption key $RK_{I \to I^*}$ for arbitrary identity I^* by just querying ReKeyGen oralce two times. The attacker can also break the IND-1PrID-CCA and IND-2PrID-CCA security of the scheme by corrupting I_{any} to get the $SK_{I_{any}}$, by associating with $RK_{I^* \to I_{any}}$. Luo et al. (2011) [24] also propose a multi-hop IBPRE scheme by using the same ReKeyGen algorithm except with $a = c, b = d$. Obviously, this multi-hop scheme can be broken by the above method. The whole attack process can be seen in Fig. 3

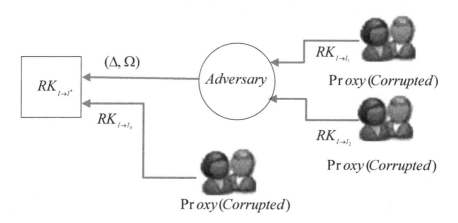

Fig. 3. Attack by using the insecure ReKeyGen algorithm

5.2   On the Transitivity Property

Luo et al. (2011) [24] pointed out their scheme has the transitivity property and discussed it to be implied by transferable property, we do not agreed with

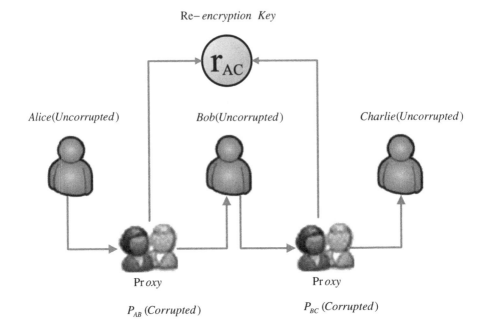

Fig. 4. Transitivity property

their conclusion. The transferable property and transitivity property are independent. They also think the transitivity property to be a useful property, but we think this property is not good for unidirectional proxy re-encryption schemes, which can be seen in Fig. 4. For example, the manager Alice delegates his reading email ability to her secretary Bob, and Bob delegates his reading email ability to his friend Carol. Now if the identity based proxy re-encryption scheme has the transitivity property, the proxies can find out the re-encryption keys from Alice to Carol, but Alice do not even know anything about Carol and Carol can read Alice's emails. Obviously, this contradict with our normal view on delegation. Furthermore, we can not easily derive the transitivity property from their schemes.

6 Conclusion

In this paper, we indicate that the proposal on multi-use CCA-secure IBPRE proposed by Wang *et al.* is not CCA-secure. We also point out another IBPRE scheme proposed by Luo et al. is not secure either. The reason for their schemes are not secure is due to the complicated security model of IBPRE or due to the insecure ReKeyGen algorithm.

Acknowledgements. An extended abstract of this paper has been published in INCoS 2012 [25]. This work was supported by the Changjiang Scholars and Innovation Research

Team in University (Grant NO. IRT 1078), the Key Problem of NFSC-Guangdong Union Foundation (Grant NO. U1135002), the Major Nature Science Foundation of China (Grant NO. 61370078), National High Technology Research and Development Program (863 Program)(No. 2015AA011704), Nature Science Foundation of China (Grant NO. 61103230, 61272492, 61202492, 61402531), Natural Science Foundation of Shaanxi Province (Grant No. 2014JM8300, 2014JQ8358, 2014JQ8307).

References

1. Blaze, M., Bleumer, G., Strauss, M.: Divertible protocols and atomic proxy cryptography. In: Nyberg, K. (ed.) EUROCRYPT 1998. LNCS, vol. 1403, pp. 127–144. Springer, Heidelberg (1998)
2. Ateniese, G., Fu, K., Green, M., Hohenberger, S.: Improved proxy re-encryption schemes with applications to secure distributed storage. ACM NDSS **2005**, 29–43 (2005)
3. Ateniese, G., Fu, K., Green, M., Hohenberger, S.: Improved proxy re-encryption schemes with applications to secure distributed storage. ACM Trans. Inf. Syst. Secur. **9**(1), 1–30 (2006)
4. Canetti, R., Hohenberger, S.: Chosen ciphertext secure proxy re-encryption. In: ACM CCS 2007, pp. 185–194 (2007)
5. Libert, B., Vergnaud, D.: Unidirectional chosen-ciphertext secure proxy re-encryption. In: Cramer, R. (ed.) PKC 2008. LNCS, vol. 4939, pp. 360–379. Springer, Heidelberg (2008)
6. Green, M., Ateniese, G.: Identity-based proxy re-encryption. In: Katz, J., Yung, M. (eds.) ACNS 2007. LNCS, vol. 4521, pp. 288–306. Springer, Heidelberg (2007)
7. Dodis, Y., Ivan, A.: Proxy cryptography revisited. In: NDSS (2003)
8. Koo, W., Hwang, J., Lee, D.: Security vulnerability in a non-interactive ID-based proxy re-encryption scheme. Inf. Process. Lett. **109**(23–24), 1260–1262 (2009)
9. Chu, C.-K., Tzeng, W.-G.: Identity-based proxy re-encryption without random oracles. In: Garay, J.A., Lenstra, A.K., Mambo, M., Peralta, R. (eds.) ISC 2007. LNCS, vol. 4779, pp. 189–202. Springer, Heidelberg (2007)
10. Matsuo, T.: Proxy re-encryption systems for identity-based encryption. In: Takagi, T., Okamoto, E., Okamoto, T., Okamoto, T. (eds.) Pairing 2007. LNCS, vol. 4575, pp. 247–267. Springer, Heidelberg (2007)
11. Wang, X., Yang, X.: On the insecurity of an identity based proxy re-encryption. Fundamental Informaticae **98**(2–3), 277–281 (2010)
12. Tang, Q., Hartel, P., Jonker, W.: Inter-domain identity-based proxy re-encryption. In: Yung, M., Liu, P., Lin, D. (eds.) Inscrypt 2008. LNCS, vol. 5487, pp. 332–347. Springer, Heidelberg (2009)
13. Ibraimi, L., Tang, Q., Hartel, P.H., Jonker, W.: A type-and-identity-based proxy re-encryption scheme and its application in healthcare. In: Jonker, W., Petković, M. (eds.) SDM 2008. LNCS, vol. 5159, pp. 185–198. Springer, Heidelberg (2008)
14. Lai, J., Zhu, W., Deng, R., Liu, S., Kou, W.: New constructions for identity-based unidirectional proxy re-encryption. J. Comput. Sci. Technol. **25**(4), 793–806 (2010)
15. Shao, J., Lu, R., Lin, X.: FINE: a fine-grained privacy-preserving location-based service framework for mobile devices. In: INFOCOM 2014, pp. 244–252 (2014)
16. Liu, X., Zhang, Y., Wang, B., Yan, J.: Mona: secure multi-owner data sharing for dynamic groups in the cloud. IEEE Trans. Parallel Distrib. Syst. **24**(6), 1182–1191 (2013)

17. Li, J., Chen, X., Li, M., Li, J., Lee, P., Lou, W.: Secure deduplication with efficient and reliable convergent key management. IEEE Trans. Parallel Distrib. Syst. **25**(6), 1615–1625 (2014)
18. Li, J., Kim, K.: Hidden attribute-based signatures without anonymity revocation. Inf. Sci. **180**(9), 1681–1689 (2010)
19. Li, J., Wang, Q., Wang, C., Ren, K.: Enhancing attribute-based encryption with attribute hierarchy. Mob. Netw. Appl. **16**(5), 553–561 (2011)
20. Li, J., Huang, X., Li, J., Chen, X., Xiang, Y.: Securely outsourcing attribute-based encryption with checkability. IEEE Trans. Parallel Distrib. Syst. **25**(8), 2201–2210 (2014)
21. la Prieta, F.D., Rodríguez, S., Bajo, J., Corchado, J.M.: +Cloud: a virtual organization of multiagent system for resource allocation into a cloud computing environment. In: Nguyen, N.T., Kowalczyk, R., Corchado, J.M., Bajo, J. (eds.) TCCI XV. LNCS, vol. 8670, pp. 164–181. Springer, Heidelberg (2014)
22. Wang, X., Ma, J., Yang, X.: A new proxy re-encryption scheme for protecting critical information systems. J. Ambient Intell. Humanized Comput. (2015). doi:10.1007/s12652-015-0261-3
23. Wang, H., Cao, Z., Wang, L.: Multi-use and unidirectional identity-based proxy re-encryption schemes. Inf. Sci. **180**(20), 4042–4059 (2010)
24. Luo, S., Shen, Q., Chen, Z.: Fully secure unidirectional identity-based proxy re-encryption. In: Kim, H. (ed.) ICISC 2011. LNCS, vol. 7259, pp. 109–126. Springer, Heidelberg (2012)
25. Zhang, J., Wang, X.: Security analysis of a multi-use identity based CCA-secure proxy re-encryption scheme. In: 2012 Fourth International Conference on Intelligent Networking and Collaborative Systems (INCos 2012), pp. 581–586 (2012)

Enabling Vehicular Data with Distributed Machine Learning

Cristian Chilipirea[1], Andreea Petre[1], Ciprian Dobre[1(✉)], Florin Pop[1], and Fatos Xhafa[2]

[1] University Politehnica of Bucharest,
Splaiul Independentei 313, Bucharest, Romania
ciprian.dobre@cs.pub.ro
[2] Universitat Politecnica de Catalunya, Girona Salgado 1-3,
08034 Barcelona, Spain

Abstract. Vehicular Data includes different facts and measurements made over a set of moving vehicles. Most of us use cars or public transportation for our work commute, daily routines and leisure. But, except of our destination, possible time of arrival and what is directly around us, we know very little about the traffic conditions in the city as a whole. Because all roads are connected in a vast network, events in other parts of town can and will directly affect us. The more we know about the traffic inside a city, the better decisions we can make. Vehicular measurements may contain a vast amount of information about the way our cities function. Information that can be used for more than improving our commute, it is indicative of other features of the city like the amount of pollution in different regions. All the information and knowledge we can extract, can be used to directly improve our life.

We live in a world where data is constantly generated and we store it and process it at an ever growing rate. Vehicular Data does not stray from this fact and is rapidly growing in size and complexity, with more and more ways to monitoring traffic, either from inside cars or from sensors placed on the road. Smartphones and in-car-computers are now common and they can produce a vast amount of data: it can identify a cars location, destination, current speed and even driving habits.

Machine learning is the perfect complement for Big Data, as large data sets can be rendered useless without methods to extract knowledge and information from them. Machine learning, currently a popular research topic, has a large number of algorithms design to achieve this task, of knowledge extraction. Most of these techniques and algorithms can be directly applied to Vehicular Data.

In this article we demonstrate how the use of a simple algorithm, k-Nearest Neighbors, can be used to extract valuable information from even a relatively small vehicular data set. Because of the vast size of most of our cities and the number of cars that are on their roads at any time of the day, standard machine learning systems do not manage to process data in a manner that would permit real time use of the extracted information. A solution to this problem is brought by distributed systems and cloud processing. By parallelizing and distributing machine learning algorithms we can use data at its highest potential and with

© Springer-Verlag Berlin Heidelberg 2015
N.T. Nguyen et al. (Eds.): Transactions on CCI XIX, LNCS 9380, pp. 89–102, 2015.
DOI: 10.1007/978-3-662-49017-4_6

little delay. Here, we show how this can be achieved by distributing the k-Nearest Neighbors machine learning algorithm over MPI. We hope this would motivate the research into other combinations of merging machine learning algorithms with Vehicular Data sets.

Keywords: Big data · Machine learning · MPI · Cloud systems · Distributed processing

1   Introduction

Vehicular data consists of all measurements that are executed and generated on the cars that participate in traffic. It can represents a vast amount of information about the cities we live in. As a society, we have reached a point of very high mobility; we all travel long distances to get to our jobs, to our schools, to our leisure activities. All this movement represents the life of a city. We do have the individual tools to measure what we are doing, what our destination is and what is the fastest way to get to it, but when we look at large groups of people participating in traffic we do not really understand what is happening, what are the common flows or patterns that people follow, how we make use of our roads. This is how we can use Vehicular Data, to understand the dynamics of our cities and to try and improve them.

The types and the complexity of data are constantly growing. Until recently only external data generators were used, like sensors on the road that would be able to detect the amount of used capacity of the road at a given time. These are complex systems that are difficult to build as well as expensive. Now, with the growth in the number of wearable devices and vehicles with on board computers, data can be generated from inside the traffic flow and at a higher velocity then we were ever able to. A simple smartphone has numerous sensors such as GPS or accelerometer, that can be used to gather data about a car: "Where is it?", "What speed does it have?" and even "Where is it going?".

With the vast number of traffic participants and the large number of people that own a smartphone, or a car capable of sending real time information about itself, Vehicular Data is growing at a large pace. Public repositories with traffic data already are available from projects such as T-Drive [22], Cabspotting [13], and taxicabs [7]. The data being collected can lead to very large volumes, making the process of extracting information out of it rather complex. Popular processing methods rely on Machine Learning algorithms. Currently an important topic of research, such algorithms provide an efficient way to analyze large amounts of data. The complexity of processes that stand behind traffic flow is so large, that only data mining algorithms - from the domains of structure mining, graph mining, data streams, large-scale and temporal data mining - may bring efficient solutions for these problems.

There is not much surprise that the topic of efficient processing of vehicular data is today quite popular with scientists and practitioners alike. Competitions like the IEEE ICDM Contest [18] are designed to ask researchers to devise the best possible algorithms that tackle problems of traffic flow prediction, for the purpose of intelligent driver navigation and improved city planning. Projects such as iDiary [14] develop advance algorithms to filter/derive/infer information out of this Big vehicular Data. However, even if the vast majority of work concentrate on designing the best solutions to deal with information extraction, little work has been done towards optimizing the data mining process itself.

In this article, our first contribution consists in showing how using a basic, simple Machine Learning algorithm, k-Nearest Neighbors, we are able to drastically improve the processing of information about traffic inside a city (Sect. 3). We next show how the ML algorithm can be distributed over multiple machines, and how it scales with the number of processors it uses (Sect. 4). We present in Sect. 5 a discussion about why we need to make this systems truly scalable and why we should allocate researching resources into these problems. Section 6 presents our conclusions.

2   Related Work

The importance and variety in uses of Vehicular Data and the use of Machine Learning to optimize traffic, is demonstrated by various authors [21,23,25]. Pau and Tse [26] present a solution where vehicles are used to measure air parameters as well as urban traffic. The data generated is then used to better understand the pollution levels in cities such as Macao. Cars represent a big polluter and the more information we have about cars the better we can understand their impact on pollution. Fu et al. [15] present an assessment of vehicular pollution using data about the estimated number of cars. This type of work can enormously benefit from accurate vehicle usage data that could be extracted from large vehicular data sets.

Safar [28] presents the use of vehicular data to answer kNN queries. The author tries to find the k nearest objects to a location on a map. In addition, vehicular data is used to improve the response time of these queries. We note this work because of the use of both vehicular data and the k-Nearest Neighbors (k-NN) algorithm (in fact, the work is among the first to make use of the k-NN algorithm in conjunction with vehicular data). However, the way in which they apply the later, and the purpose, are vastly different than our work.

Processing of vehicular data is commonly done with statistical techniques. Williams and Hoel [31] present a solution where the data is provided by a road control agency (the highway agency), and the processing is designed to extract weekly or seasonal patterns in the usage of the analyzed highways.

Old systems (see Hsieh et al. [17] or Coifman et al. [11]) use video feeds to do vehicle tracking and classification. Such systems can still be used as Vehicular data generators, but results show they fail to scale well and/or incur high/prohibitive costs for large-scale adoption (i.e., a lot of cameras need to be placed around the city, and processing video feeds requires powerful specialized equipment). Zhu et al. [33] present a system that uses RFID tags on vehicles, and a city wide network is designed to gather vehicular data. Even though this system is less expensive and more scalable, it still requires large investments in the network and the willingness of the tracked vehicles to install an RFID tag on their cars. Systems later evolved to transmit real time data from inside a vehicle. Chadil et al. [9] present a system that uses a custom board with GPS and GPRS to enable vehicle localization and tracking.

Even though vehicular data is very noisy, current research is trying to improve its quality. Schubert et al. [29] present a comparison between the models used to increase the accuracy of location data. With the high error rate of GPS receivers, and the importance of geospatial localization for vehicular data, this type of research is vital for the improvement of such data sets. Brakatsoulas et al. [8] take a different approach and try to match GPS data to a street map, doing all the necessary corrections in the process.

The use of machine learning techniques over Vehicular Data has been done in works like [1], where this method is used to improve traffic signal control systems by making these systems truly adaptive. However the dataset used is considerably smaller, it covers only one intersection, while we are looking at city wide data sets.

Similar to our work, Sun et al. [30] present the use of Bayesian networks to achieve traffic flow forecasting. Unlike our work, their test set is provided by the Traffic Management Bureau of Beijing and it is given in vehicles per hour, a metric that is not always available. Because of this less noisy data set and its size there was also no need to improve the execution time of their solution.

The need for scalability in the processing of vehicular data is also identified by Biem et al. [6], where they use a small compute cluster to process GPS data from taxis and trucks. This data is processed as streams and it is used to create real-time speed maps over the city.

Zhang et al. [32] present a way to distribute kNN Joins over MapReduce. Unlike our solution, their work concentrates on Joins on extremely large data sets, the main algorithm is however similar and they do obtain a speedup of 4 with 20 reducers. A different solution to the same problem of kNN Joins is presented by Lu et al. [20]. The authors also chose to distribute the computation using the Hadoop MapReduce framework. Another article discussing machine learning algorithms over Hadoop is [16], where multiple machine learning algorithms are tested over the Hadoop framework. Similarly, authors in [19] present a similar solution, Distributed GraphLab, aimed for Cloud applications.

Reducing the execution time of k-nearest neighbors has also been done by parallelizing it using GPU cores [5]. Their solution can be used in combination with the solution we present in this paper to achieve even greater processing speed boosts.

3   Application of ML Algorithms on Traffic Data

The traces we use were obtained from the CRAWDAD public repository [12], a community resource for archiving wireless data. The website contains a large number of traces that can be used in Mobile Ad-Hoc Networking and Vehicular Mobile Ad-Hoc Networking simulations. We selected two popular vehicular traces to run our experiments on:

- Roma trace [2];
- San Francisco epfl trac [27].

Table 1 includes the main characteristics of these two traces. San Francisco is the largest one, with double the number of data items and more than half the number of taxis being recorded. The traces themselves have also been taken six years apart, with the more recent one being the San Francisco trace.

Table 1. Trace characteristics.

	Roma	San Francisco
of cars	316	536
of data items	11.219.955	21.673.309
Start Date	17-May-2008	01-February-2014
End Date	10-June-2008	02-March-2014

Both trace contains location coordinates of taxis (in a format similar to *taxi_id*, *timestamp* and *location* (*latitude* and *longitude*)). We chose these two data sets because they span over a moderately large time period, and they both include a large number of entries (an aspect particularly important to evaluate scalability aspects).

We believe that in any type of long time measurement regarding human activity one should be able to distinguish a day-night pattern. This pattern has a few main characteristics. First, it is an excellent way to validate the data, and a lack of such a pattern might indicate that something is wrong with the data generation/recording process or that the data measures something that is not directly affected by human behavior.

The data was formatted to take the following header: *time_of_day* (broken in 30 min intervals); *day_of_week*; *speed*. The *id* field was not relevant for our experiments. We also removed all the data points where speed was lower than 0.5 km/h for a longer time period (to exclude parked vehicles from our measurements).

The newly formatted data was then processed using the *k-Nearest Neighbors* (k-NN) algorithm. The entire data set was used as a training set, and for evaluation we created a secondary set with 48 *time_of_day* intervals and null speed values. The predicted results, indicative of the k-NN model, are visible in Figs. 1 and 2. In the figure we represented the mean speed computed for the various times of day (this is much lower than the maximum permitted speed, as there are always cars driving slowly, as expected in congested cities). Furthermore, the traces contain data about monitored taxis, which have an unusual driving behavior compared to normal cars. For example, they slow down when they want to search for a customer or for a building from which they received a request.

In the same traffic conditions we expect that normal cars would introduce a higher mean speed. But there are still a lot of traffic events that lower the mean speed: cars have to stop at cross-roads, cars stop to pick up other passengers, and cars slow down when the driver is searching for a location, emergency vehicles force drivers to slow down or stop to give priority. All these lower the mean speed of cars within a city.

For all processing and all graphs we used a k value of 500, and the k-Nearest Neighbor algorithm was set to calculate the mean speed over all the neighbor values.

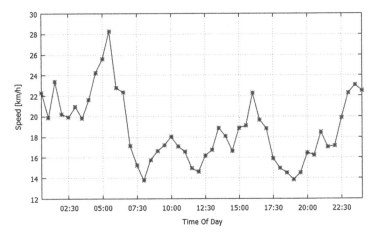

Fig. 1. k-Nearest Neighbors model for speed, Roma.

Figure 1 present the average traveling speed per time of day, obtained for the Roma trace. We can discern a day-night pattern in the obtained speed model. During the day, the speed lowers considerably compared to nighttime. This happens mainly because of high traffic during the day, with a lot of cars stuck in traffic – the mean speed is lowered for all the cars that participate in said traffic. This model is also a good predictor of what mean speeds to expect at a certain hour, or when it is best to make a trip so that you can achieve maximum speed. The figure is averaged for the entire city, but using the same principle, we can and did obtain data for different parts of the city, or even with fine granularity we can reach the street level.

The San Francisco model (Fig. 2), shows a similar day-night pattern. This model also gave us some previously unexpected information. We are able to clearly identify the "rush hours", moments during the day where there are so many cars in traffic that the mean speed is lowered by a large factor. In the mode we can observe both the morning and the afternoon "rush hour" moments, where the mean speed is at its minimum. After a closer inspection of the Roma model, we identified the same drops in mean speed at similar time intervals.

What we did not expect was that when we overlap the two models, we would find a relatively close correlation between them. In Fig. 3 we plotted the over-lapped data. Here, most of the raises and drops in the mean speed almost align.

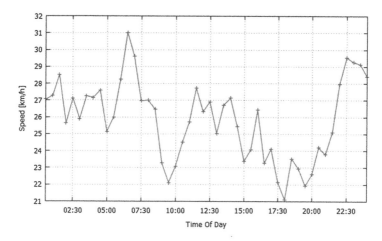

Fig. 2. k-Nearest Neighbors model for speed, San Francisco.

Considering the day-night cycle in mean speed is a sort of a "heart-beat" of the city, we find it extremely interesting that two very different cities in different parts of the world with different cultures, and data gathered years apart, have such a similar "heart-beat".

We note that the mean speed over the entire day in San Francisco is higher than the mean speed of the Roma trace. We believe this to caused because of geography; the city of Roma has hills, unlike San Francisco, and because it is an older city, dating back to the start of the Roman empire, the streets are not as wide, permitting less traffic, at lower speeds.

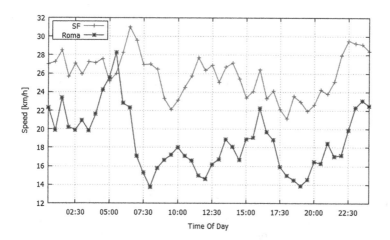

Fig. 3. k-NN San Francisco/Roma model comparison.

Fig. 4. The city of Roma as obtained from Open Street Maps [24].

For the Roma trace we made another analysis. We wanted to extract density maps and see how these change as effect of the day-night difference.

We divided the map into 10.000, or 100 by 100, sub-regions. Then, for each sub-region, we counted the number of cars between the hours 01:00 and 07:00 for the night, and between the hours 14:00 and 20:00 for the day, for each day individually. This data was then processed using the k-NN algorithm to extract a mean value for each sub region. The results can be seen in Figs. 5 and 6. In these figures, a black color means a high number of cars, while a white color means a smaller number of cars, even 0.

It is very clear that during the day the car density is higher in the center of the city. By comparing these images with Fig. 4, the map of the city, we can see how the cars follow the main city roads. We note that using this analysis we were able to map most of the roads in Roma using only GPS data from taxis.

4   Distributing k-Nearest Neighbors

Running k-Nearest Neighbors over multiple machines can be done in multiple ways. We chose to use the Massage Passing Interface (MPI) framework. MPI is a distributed framework (and a message passing library interface speciïïñAcation for parallel programming) mostly used in high performance computing clusters.

MPI is appropriate for data-dependent iterative algorithms based on message passing. People have long opposed MPI to MapReduce (the next widest used model/framework for parallel/distributed ML algorithms – Chen et al. [10] make

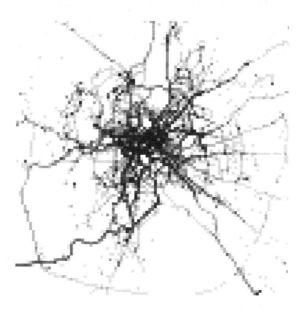

Fig. 5. Car density during the night.

Fig. 6. Car density during the day.

an advanced comparison of the two), which is most suitable for non-iterative algorithms without lots of data exchanges. We chose MPI as an alternative to what we found in the literature about distributing k-NN using the Hadoop MapReduce framework (see Zhang et al. [32], Lu et al. [20], or Gillick et al. [16]). To the best of our knowledge, today there is no MPI implementation of k-NN available.

Our solution reads both the training set and the test set in the first task ($rank == 0$), and this data is spread to all other tasks running on the distributed machines. Each machine then processes its part of the test set.

In k-Nearest Neighbors each element in the test set needs to be compared with all elements in the training set to determine: which are the k-nearest neighbors. After all the tasks are finished processing their part of the test set, the predicted value is calculated for each point in the data set and the data is gathered to the first task.

During our experiments we used different values for k, and we split the Roma test set into 80 % for the training set and 20 % for the test set.

The tests were run on a cluster of IBM Xeon E5630 rated at 2.53 GHz with 32 GB or RAM. During the experiments we used between 1 and 10 different such machines. We note here that each experiment was run 3 times and the results were very similar, in the graph we show the mean execution time between the 3 different runs.

In Fig. 7 we can observe the execution time for the k-NN algorithm on the Roma data set. We used k values of 100, 300, 500, 700 and 1000.

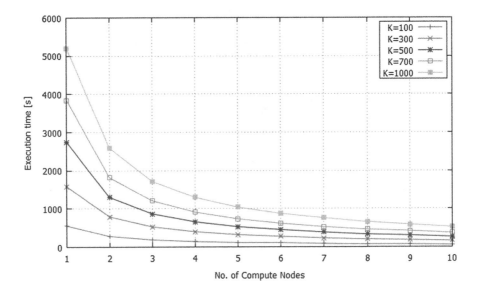

Fig. 7. Execution Time for distributed k-NN.

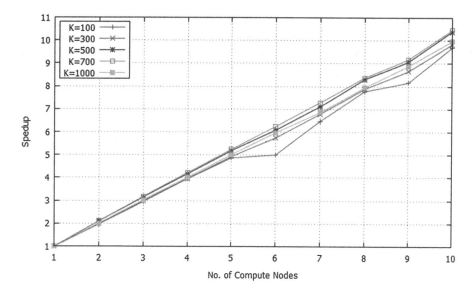

Fig. 8. Speedup for distributed k-NN.

The speedup for our implementation of the algorithm is almost linear with the number of compute nodes. This is shown in Fig. 8, which shows the speedup for all the values of k we used in our experiment. For smaller values of k, the speedup is insignificant (this is visible for k = 100, and degrades for lower values).

5   Discussion

Each experiment in this paper took between 30 min to 1 h to execute on one single core machine, excluding data parsing or formatting. The traces contained data for just 1 month for over 500 cars. When we are looking at scaling these solution to city size we consider that every car is a potential data generator. This means millions of cars per city each generating one line of data every few seconds. To make things even worse, some machine learning algorithms, such as neural networks or convolution neural networks, have even higher execution times than the one we presented, but they could be used to extract even more interesting information from the data sets and to make more accurate predictions.

With the extremely large number of data generators and the high execution time for machine learning algorithms building a complex real time solution can prove difficult. It is important to find the best ways to use all the resources provided by a modern computers, executing code on multiple cores, using both CPU and GPU, as well as enable the algorithms to make use of multiple machines.

Another factor that can raise the complexity of future Intelligent Transportation System (ITS) applications is the complexity of the generated data itself. Soon only timestamps and GPS coordinates will not be enough, and more complex sensors will provide a large variety of data. With the fast integration of smartphones that have powerful photo and video cameras, it is no longer far fetch to consider that soon one system will have to process video data generated by all the cars inside a city.

6   Conclusion

In this article we presented our analysis over two distinct vehicular traces, one for the city or Roma and another for San Francisco, using the k-Nearest Neighbor algorithm. We chose to use the k-Nearest Neighbors algorithm because it is one of the simplest (and yet powerful) machine learning tool.

With the analysis we showed that with the use of machine learning algorithms important information can be extracted from even basic vehicular data sets. The used data sets only contain time stamp and GPS coordinates (latitude and longitude) but other data sets may contain data from other sensors such as microphones, to measure noise levels or chemical sensors that could measure the level of pollution. With the minimal available data we manage to identify a day night cycle pattern, rush hours and we manage to build a model that could be used to predict high traffic intervals and regions with high vehicular density.

Then we showed how machine learning algorithms such as k-Nearest Neighbors can exploit the power of compute clusters or even clouds by reducing execution time through distributed processing. We achieve this using the popular MPI framework.

The need for highly scalable machine learning algorithms for use with vehicular data is further explored in our discussion section.

As future work we believe more machine learning algorithms should be tested in combination with vehicular data. We believe that with the right combination of data and algorithm surprising and interesting information could be extracted.

More machine learning algorithms should be parallelized or distributed so that they can be efficiently executed over multiple machines. A lot of work in this direction has been done in projects such as Mahout [3] or Spark MLLib [4], but not all machine learning algorithms have been integrated into these package. For instance *neither contain a k-Nearest Neighbors implementation*.

More vehicular traces need to be generated and analyzed. The vehicular traces we used only contained data from taxis, which made them slightly biased. A vehicular trace that contains all kinds of vehicles would be extremely interesting for the research community. It would also be fascinating to have public traces that contain other sensor data such as pollution or noise levels.

Acknowledgment. This work was supported by the Romanian national project Mobi-Way, Project PN-II-PT-PCCA-2013-4-0321. The authors would like to thank reviewers for their constructive comments and valuable insights.

References

1. Abdulhai, B., Pringle, R., Karakoulas, G.J.: Reinforcement learning for true adaptive traffic signal control. J. Transp. Eng. **129**(3), 278–285 (2003)
2. Amici, R., Bonola, M., Bracciale, L., Rabuffi, A., Loreti, P., Bianchi, G.: Performance assessment of an epidemic protocol in vanet using real traces. Procedia Comput. Sci. **40**, 92–99 (2014)
3. Apache: Mahout (2015). https://mahout.apache.org/
4. Apache: Spark mllib (2015). https://spark.apache.org/mllib/
5. Barrientos, R., Gómez, J., Tenllado, C., Prieto, M.: Heap based k-nearest neighbor search on gpus. In: Congreso Espanol de Informática (CEDI), pp. 559–566 (2010)
6. Biem, A., Bouillet, E., Feng, H., Ranganathan, A., Riabov, A., Verscheure, O., Koutsopoulos, H., Moran, C.: Ibm infosphere streams for scalable, real-time, intelligent transportation services. In: Proceedings of the 2010 ACM SIGMOD International Conference on Management of Data, pp. 1093–1104. ACM (2010)
7. Bracciale, L., Bonola, M., Loreti, P., Bianchi, G., Amici, R., Rabuffi, A.: CRAWDAD data set roma/taxi (v. 2014-07-17), Jul 2014. Downloaded from http://crawdad.org/roma/taxi/
8. Brakatsoulas, S., Pfoser, D., Salas, R., Wenk, C.: On map-matching vehicle tracking data. In: Proceedings of the 31st International Conference on Very Large Data Bases, pp. 853–864. VLDB Endowment (2005)
9. Chadil, N., Russameesawang, A., Keeratiwintakorn, P.: Real-time tracking management system using gps, gprs and google earth. In: 5th International Conference on Electrical Engineering/Electronics, Computer, Telecommunications and Information Technology, ECTI-CON 2008, vol. 1, pp. 393–396. IEEE (2008)
10. Chen, W.Y., Song, Y., Bai, H., Lin, C.J., Chang, E.Y.: Parallel spectral clustering in distributed systems. IEEE Trans. Pattern Anal. Mach. Intell. **33**(3), 568–586 (2011)
11. Coifman, B., Beymer, D., McLauchlan, P., Malik, J.: A real-time computer vision system for vehicle tracking and traffic surveillance. Transp. Res. Part C: Emerg. Technol. **6**(4), 271–288 (1998)
12. Dartmouth: Crowdad (2015). http://crawdad.org/
13. exploratorium: Cabspotting (2015). http://cabspotting.org/index.html
14. Feldman, D., Sugaya, A., Sung, C., Rus, D.: idiary: From gps signals to a text-searchable diary. In: Proceedings of the 11th ACM Conference on Embedded Networked Sensor Systems, p. 6. ACM (2013)
15. Fu, L., Hao, J., He, D., He, K., Li, P.: Assessment of vehicular pollution in china. J. Air Waste Manage. Assoc. **51**(5), 658–668 (2001)
16. Gillick, D., Faria, A., DeNero, J.: Mapreduce: Distributed computing for machine learning. Berkley, 18 Dec 2006
17. Hsieh, J.W., Yu, S.H., Chen, Y.S., Hu, W.F.: Automatic traffic surveillance system for vehicle tracking and classification. IEEE Trans. Intell. Transp. Syst. **7**(2), 175–187 (2006)
18. IEEE, TomTom: Ieee icdm contest: Tomtom traffic prediction for intelligent gps navigation (2010). http://tunedit.org/challenge/IEEE-ICDM-2010
19. Low, Y., Bickson, D., Gonzalez, J., Guestrin, C., Kyrola, A., Hellerstein, J.M.: Distributed graphlab: a framework for machine learning and data mining in the cloud. Proc. VLDB Endowment **5**(8), 716–727 (2012)
20. Lu, W., Shen, Y., Chen, S., Ooi, B.C.: Efficient processing of k nearest neighbor joins using mapreduce. Proc. VLDB Endowment **5**(10), 1016–1027 (2012)

21. Mavromoustakis, C.X., Kormentzas, G., Mastorakis, G., Bourdena, A., Pallis, E., Rodrigues, J.: Context-oriented opportunistic cloud offload processing for energy conservation in wireless devices. In: Globecom Workshops (GC Wkshps), pp. 24–30. IEEE (2014)

22. Microsoft, R.: T-drive: Driving directions based on taxi traces (2015). http://research.microsoft.com/en-us/projects/tdrive/

23. Mousicou, P., Mavromoustakis, C.X., Bourdena, A., Mastorakis, G., Pallis, E.: Performance evaluation of dynamic cloud resource migration based on temporal and capacity-aware policy for efficient resource sharing. In: Proceedings of the 2nd ACM Workshop on High Performance Mobile Opportunistic Systems, pp. 59–66. ACM (2013)

24. OpenStreetMap: Openstreetmap (2015). https://www.openstreetmap.org

25. Papadakis, S.E., Stykas, V., Mastorakis, G., Mavromoustakis, C.X., et al.: A hyperbox approach using relational databases for large scale machine learning. In: 2014 International Conference on Telecommunications and Multimedia (TEMU), pp. 69–73. IEEE (2014)

26. Pau, G., Tse, R.: Challenges and opportunities in immersive vehicular sensing: lessons from urban deployments. Sig. Process. Image Commun. **27**(8), 900–908 (2012)

27. Piórkowski, M., Sarafijanovic-Djukic, N., Grossglauser, M.: A parsimonious model of mobile partitioned networks with clustering. In: First International Communication Systems and Networks and Workshops, COMSNETS 2009, pp. 1–10. IEEE (2009)

28. Safar, M.: K nearest neighbor search in navigation systems. Mob. Inf. Syst. **1**(3), 207–224 (2005)

29. Schubert, R., Richter, E., Wanielik, G.: Comparison and evaluation of advanced motion models for vehicle tracking. In: 2008 11th International Conference on Information Fusion, pp. 1–6. IEEE (2008)

30. Sun, S., Zhang, C., Yu, G.: A bayesian network approach to traffic flow forecasting. IEEE Trans. Intell. Transp. Syst. **7**(1), 124–132 (2006)

31. Williams, B.M., Hoel, L.A.: Modeling and forecasting vehicular traffic flow as a seasonal arima process: Theoretical basis and empirical results. J. Transp. Eng. **129**(6), 664–672 (2003)

32. Zhang, C., Li, F., Jestes, J.: Efficient parallel knn joins for large data in mapreduce. In: Proceedings of the 15th International Conference on Extending Database Technology, pp. 38–49. ACM (2012)

33. Zhu, H., Zhu, Y., Li, M., Ni, L.M.: Hero: Online real-time vehicle tracking in shanghai. In: IEEE The 27th Conference on Computer Communications, INFOCOM 2008. IEEE (2008)

Adapting Distributed Evolutionary Algorithms to Heterogeneous Hardware

Carolina Salto[1]($\boxtimes$) and Enrique Alba[2]

[1] Universidad Nacional de La Pampa - CONICET, General Pico, Argentina
saltoc@ing.unlpam.edu.ar
[2] Universidad de Málaga, Málaga, Spain
eat@lcc.uma.es

Abstract. Distributed computing environments are nowadays composed of many heterogeneous computers able to work cooperatively. Despite this, the most of the work in parallel metaheuristics assumes a homogeneous hardware as the underlying platform. In this work we provide a methodology that enables a distributed genetic algorithm to be customized for higher efficiency on any available hardware resources with different computing power, all of them collaborating to solve the same problem. We analyze the impact of heterogeneity in the resulting performance of a parallel metaheuristic and also its efficiency in time. Our conclusion is that the solution quality is comparable to that achieved by using a theoretically faster homogeneous platform, the traditional environment to execute this kind of algorithms, but an interesting finding is that those solutions are found with a lower numerical effort and even in lower running times in some cases.

1 Introduction

Parallel and distributed computing environments became popular in the past decades as a way to provide the needed computing power to solve complex problems, representing an effective strategy for the execution of distributed evolutionary algorithms (dEAs) [27]. Most of the reported results on dEAs assume that the underlying computing environments have identical features (homogeneous environment) regarding not only hardware (processors, memory, network) but also software (operating system) components [2,20]. This kind of hardware homogeneity is increasingly difficult to find in modern labs. It is quite hard to maintain a cluster of similar processors along a period of time, because of their failures and the new and different hardware replacing them. Furthermore, the rapid development of technology in designing processors, networks, and data storage devices together with the constantly decreasing ratio between cost and performance allow researchers to use new up-to-date computational resources. As a consequence, the coexistence of new and old equipment in a computing environment has shown the grow of heterogeneous parallel platforms, which are nowadays very common in any laboratory, company, and research institution.

© Springer-Verlag Berlin Heidelberg 2015
N.T. Nguyen et al. (Eds.): Transactions on CCI XIX, LNCS 9380, pp. 103–125, 2015.
DOI: 10.1007/978-3-662-49017-4_7

Despite the widespread scenario from the point of view of heterogeneous architectures, the field of metaheuristic algorithms that exploits the heterogeneous architectures in a especialized way has been seldom addressed. A seminal work dealing with heterogeneous computing environments and dEAs can be found in [3]. More recent works about heterogeneous environments can also be found in [6,12,15,17,22,23]. These works were focused on solving a given problem, and not much in building a methodology that researchers could use when facing heterogeneous settlements. In this sense, in [12] authors made an original advance in the proposal of a general model to design heterogeneous algorithms depending on the underlying heterogeneous platform.

In the present article we propose a new methodological procedure, and a subsequent algorithmic design, to deal with heterogeneous parallel environments. We called it HAPA: *Hardware Aware Parallel Algorithms* methodology. The goal is to guide an efficient and numerically accurate deployment of a metaheuristic like a dEA or a dGA (distributed genetic algorithm) onto a set of machines with processors running at different clock speeds, dissimilar principal memory capacities, and operating systems. The dGA considered in this work is a multi-population (island) model which performs sparse exchanges of individuals (migration) between the component subpopulations. In short, we address two interesting research questions:

Q1. Can we build an algorithm using the HAPA methodology with a final accuracy comparable to that of existing algorithms running on faster homogeneous platforms?

Q2. Does the use of heterogeneous hardware allow to solve the problem in competitive execution times?

The methodology devised in this work is targeted to address these two questions (real challenges), and it can be summarized as follows. In a first phase, HAPA will analyze the heterogeneous hardware with several different benchmark programs and with the running times obtained from the dGA. These two different measures will allow us to obtain a quantitative measure for the speed of the different hardware involved. This quantitative value is obtained following a well defined methodology, what represents a deeper contribution than only the value itself. With this information, in a second phase, we will develop a novel mechanism to be used in the design of a dGA, engineered to get profit from a computing platform composed of both new and old computational equipment. Finally, in order to answer the two previous questions about the behaviour of the algorithms using HAPA, we have compared a dGA using our proposal against a traditional dGA executed on a homogeneous environment. This constitutes the third phase of the methodology related to the validation of the results.

After using HAPA we hope to be able to report that a dGA using our proposal can obtain similar hit rates as a dGA running over homogeneous hardware. We will show this in this work, as well as we will report on a reduction in the number of function evaluations needed by the new techniques, thus reinforcing the idea of the usefulness of HAPA in the design of competent dGA families of algorithms.

The remainder of this article is structured as follows. Section 2 provides a brief review of the literature dealing with dGAs and heterogeneous hardware. Section 3 presents the HAPA methodology. Section 4 introduces the test problem and the parameterizations used in the experimentation. Section 5 is devoted to describe the heterogeneous hardware. Section 6 provides two possible instantiations in the design of a dGA. Section 7 presents and examines the results validating our proposal. Section 8 summarizes our conclusions and sketches our future work.

2   Background

Let us suppose a company or laboratory has bought a cluster, composed of workstations interconnected by a communication network. Any (even new) cluster of computers will become old with time, and possibly heterogeneous due to changes in its components (partial memory or CPU updates, for example). These components are replaced with different (possibly more powerful) ones. In essence, the cluster is only homogeneous (if at all) when first installed. Additionally, any necessary increment in performance or capacity is usually achieved by replacing old/broken components with more powerful ones. This leads to the coexistence of "leftovers" from the previous installation and "new-comers" that are recently purchased, leading to the emergence of a heterogeneous computing environment in terms of performance and capacity. A situation as the one previously described is sketched in Fig. 1.

There exist only a few works proposing new algorithms for heterogeneous platforms, and none on developing a methodology to do so. A seminal proposal concerning the heterogeneous execution of parallel metaheuristics was proposed by Alba et al. [3]. In this work, the authors analyzed the way in which heterogeneous environments affect the genetic search to solve a problem, reporting

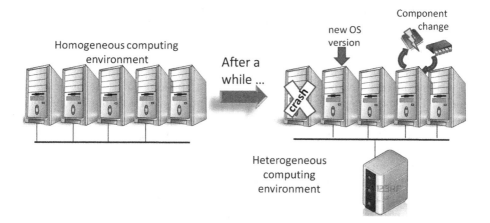

Fig. 1. Heterogeneous computing platform

a very significant reduction in the number of steps needed to solve the problem when using heterogeneous hardware. In another interesting work, Chong [9] analyzed the impact of the asynchronous and synchronous communication on a heterogeneous hardware platform and concluded that communication should be non-blocking (i.e. asynchronous) and buffered, a result that has been also confirmed in [4, 6, 19].

Branke et al. [8] considered an island model targeted at heterogeneous computer grids and examined different aspects of migration, like the connectivity pattern or the time for migration. They experimented with different ways of sorting the islands: a random sorting of the heterogeneous processors on the ring topology and a minimum and maximum difference sum sorting of the processor velocity. They compared the performance of the standard island model on the homogeneous and the heterogeneous network. They simulated the underlying computer network. Their conclusion was that the result of sorting the computers appropriately in the ring structure is competitive to homogeneous networks. Another significant result was that convergence-based migration leads to a further significant improvement both in homogeneous and heterogeneous environments. Another proposal in this line is the work of Gong et al. [17], which also analyzed the influence of different arrangements of heterogeneous computing resources in the execution of dGAs.

A parallel genetic algorithm with a hierarchical distribution was presented in [18], developed using heterogeneous grid computing environments. In each cluster a subpopulation is evolved while the chromosome evaluation is carried out in a different node of the cluster. A theoretical analysis of the speed-up was presented. The empirical study was oriented to analyse the behavior of the algorithm under diverse grid environments having different communication protocols, cluster sizes, computing nodes, and geographically disparate clusters. The authors showed that speed-up can be attained. García et al. [15] and Meri et al. [22] considered the use of free cloud storage services to communicate a pool of distributed island running in a heterogeneous platform.

Bazterra et al. [7] defined performance metrics to understand the parallel efficiency of an algorithm running on heterogeneous systems. They proposed an adaptive parallel genetic algorithm which consists in a client-server model for heterogeneous environments. The server node evolves the population and assigns the evaluation of individuals at each processor depending of their processing velocity. They evaluated the performance of the proposal in a homogeneous and heterogeneous environment, achieving a higher efficiency in the last one. In this line of client-server models, Mostaghim [23] proposed a hybrid method using Multi-objective Particle Swarm optimization and Binary search methods for a multi-objective optimization task independent from the speed of the processors.

More recently, Dominguez and Alba [11, 12] proposed Ethane and HydroCM, two new heterogeneous parallel search algorithms specifically designed for their execution in a heterogeneous hardware platform. The proposed search algorithms, inspired by the chemical structures of ethane and hydrocarbons, were based on genetic algorithms (GA) and simulated annealing (SA). The objective was to give a general kind of parallel search technique that could later be customized to be used with different behavioral algorithms depending on the underlying hardware architecture. The reported results have shown that Ethane and HydroCM can perform better in terms of time and numerical effort when run in heterogeneous software/hardware systems than the component algorithms. Thus, it seems clear that this topic deserves more research since it is both interesting and not well-known at present.

3   The HAPA Methodology

In this section we present the basics of the proposed methodology. HAPA is a methodology in which a distributed population-based metaheuristic can deal with the differences between relative clock speeds of the processors present in a heterogeneous platform. Our methodology consists in computing (once) a ranking of processors in an offline fashion, plus an online use of this information inside the running distributed algorithm in some way (like defining new stopping conditions). This also goes in the sense of "measure once, use many" that can help modern labs to better build algorithms and better use their hardware at the same time. The HAPA methodology comprises three phases:

- Phase 1: Know your Platform.
- Phase 2: Design your Algorithm.
- Phase 3: Get the Results.

Phase 1 numerically describes the heterogeneous computing platform used in the work, with the aim of finding a ranking of processors. Consequently, this phase involves the computation of the relative differences in the velocity of each machine, taking the fastest one as a reference point. For this purpose, we use two different measures to evaluate the performance of machines: the scores from a standard scientific benchmarking software and a fresh ranking coming from the actual execution of a traditional dGA. The rationale behind this is that traditional benchmarks, although useful, are designed to run a set of operations that could not be fully representative of the operations performed in metaheuristics [5]. The objective is to determine which traditional benchmark software, if any, is able to rank the processors in the same order as that the actual dGA running benchmarking, and to use that information for the application of the HAPA methodology. This produces information to numerically know the hardware. Once a ranking is established, a relative velocity factor between each processor B and the fastest processor A can be obtained, denoted as $VF(B)$.

With the ranking obtained in the previous step, Phase 2 consists in the algorithm design of the dGA to deal with the heterogenous platform and to present the HAPA methodology. This online phase is carried out by all the component islands in an organized way. In a first step, each island i of our dGA using the HAPA methodology asks for the features of the processor where it was launched (let us say processor B), obtaining the $VF(B)$ value. Let us suppose, then, the considered parameter value for the island running in the fastest processor is set to X. Therefore, in a second step, each island i proceeds to locally compute the parameter value to a value equal to X divided by $VF(B)$; in that way the islands are coordinating the parameter values depending on the features of the processor where each one was launched. So, the methodology helps to have a more informed decision-making of the distributed algorithm's parameter values in each subpopulation, such as the total number of generations, the migration frequency, and so on. This methodology also prevents the situation that one of the islands is doing most of the work if there is a CPU much faster than the others.

Finally, Phase 3 consists in the evaluation of the proposed dGAs using the HAPA methodology to validate the assumptions made in the Introduction related to their comparative performance with respect to dGAs running in homogeneous platforms regarding final accuracy and execution times.

In Sect. 6, we explain two possible instantiations of the proposed methodology in the design of a dGA. In Sect. 6.1 we describe an application of the HAPA methodology to set the number of generations used as stop condition on each island, while in Sect. 6.2 we explain the use of HAPA to define the migration frequency. In both cases, the aim is a meaningful dynamic determination of the parameter values to profit from the differences in the hardware involved to build a more efficient/accurate algorithm. The two algorithms derived from HAPA set one parameter at a time, in order to identify the reasons of possible improvements in their performance.

4   Experimental Setup

In this section we present the necessary information to reproduce the experiments that have been carried out in this article. First we will introduce the problem used to assess the performance of our proposal: the Knapsack Problem (KP), a classical combinatorial optimization problem. In the present study we are not focusing on the solution of a particular problem (many different and specific heuristics exist for this [26]), but our aim is simply to use it to evaluate our proposals. Second, we will justify the parameters that our dGA will use.

4.1   The Knapsack Problem

The Knapsack Problem (KP) belongs to the class of NP-hard problems [16]. Given a knapsack capacity C, and a set N of n items with associated profit $p_i > 0$ and weight $w_i > 0$, the goal is to choose a subset of items such that maximizes the total profit keeping the total weight below the C capacity of the knapsack. We may assume that $w_i < C$, for $i = 1, \ldots, n$ to ensure that each item considered fits into the knapsack, and that the total weight of all the items exceeds C to avoid trivial solutions. The KP can be formulated as an integer programming model as presented in Eq. 1, where x_i is the binary decision variable of the problem that indicates whether the item i is included or not in the knapsack.

Table 1. KP instances

Instance	n	R	C	Optimal profit
KP1-1k	100	1000	1001	9147
KP1-10k	100	10000	10001	81021
KP2-1k	200	1000	1001	11238
KP2-10k	200	10000	10001	106285
KP3-1k	300	1000	1001	13643
KP3-10k	300	10000	10001	129441
KP4-1k	400	1000	1001	15939
KP4-10k	400	10000	10001	141774

$$\text{maximize} \quad \sum_{i=1}^{n} p_i x_i$$

$$\text{subject to :} \quad \sum_{i=1}^{n} w_i x_i \leq C, \qquad (1)$$

$$x_i \in \{0,1\}, \ \forall i = 1, \ldots, n$$

Four randomly generated data instances are considered as listed in Table 1, with n varying from 100 to 400 items and with two different C capacities (1001 and 10001). These instances were obtained using the generator described in [25] choosing the uncorrelated data instances type, i.e., p_j and w_j which are randomly distributed in $[1, \ldots, R]$ (no correlation between the weight and the profit of an item, in order to make the problem harder). The optimal solution of each instance (reported in Table 1) was found using the Minknap algorithm [24], an exact method based on dynamic programming.

4.2 Parameters

In our experiments, the global pool of solutions of the dGA is set to 512 solutions, which are organized into 8 islands of 64 solutions each. The tentative solutions for the KP are encoded as binary strings. The genetic operators are: binary tournament selection, two point crossover, and bit flip mutation. The crossover and mutation rates are 0.65 and $1/n$ (where n is the length of the solutions), respectively. Proportional selection is used to build up the next population from the set of parent and offspring solutions. The base migration frequency is set to 128 generations. A copy of the best individual of each subpopulation is sent and replaces the worst solution on the target island, only if it is better (only one individual is sent in each exchange). We would like to remind that the communication between islands is entirely asynchronous, so there are no sync points in the migration operation. The topology follows a unidirectional ring communication pattern. Table 2 summarizes the parameters used in the experimentation.

Table 2. Experimental parameters of all dGAs

Population size	512 individuals
Number of islands	8 islands
Selection of parents	Binary tournament
Recombination	two-point, $pc = 0.65$
Bit mutation	Bit-flip, $pm = 1/n$
Replacement	Rep_better
Migration frequency	128 generations

The code was developed using MALLBA [14], a C++ software library foster-ing rapid prototyping of hybrid and parallel algorithms, running under Linux. The considered hardware resources are the ones shown in Table 3 which are described in the next section.

Due to the stochastic nature of the algorithms, the final results are obtained after averaging the running times of 30 independent runs. A statistical analysis has been performed in order to provide the results with statistical confidence and, therefore, obtain meaningful conclusions. We use the non-parametric Kruskal-Wallis test, to distinguish meaningful differences between the mean results of all algorithm. We have considered a level of significance of $\alpha = 0.05$, in order to indicate a 95 % confidence level in the results.

5   HAPA Phase 1: Know Your Platform

We now proceed to explain the first phase of the HAPA methodology (offline phase). It consists of three parts: getting general knowledge on the platform, fine tuning this knowledge for the class of algorithms we are interested in, and a final third part in which we summarize all this knowledge into a mathematical function to be able of designing algorithms based on it. In consequence, Sect. 5.1 introduces a characterization of the heterogeneous hardware. Section 5.2 shows the scores from a standard scientific benchmarking software to get a general knowledge, while Sect. 5.3 presents a ranking coming from the actual execu-tion of a traditional dGA, the class of algorithms we are interested in. Once the machine performance has been obtained (i.e. the hardware has been trans-formed into numerical knowledge), the last Sect. 5.4 presents how to obtain the relative velocity factor between each processor and the fastest one by applying a mathematical function; this factor is the basis for the application of the HAPA methodology (online phase).

5.1   Hardware Description

The heterogeneous computing system consists of a wide range of diverse CPUs belonging to different families of processors, including single and multicore, sin-gle and multithreaded, mono and multiprocessors, 32 and 64 bits, and different

Table 3. Heterogeneous computing environment

Name	Features	RAM(GB)	#cores	#nodes	Year
CPU1	AMD Athlon XP2000+ at 1.67 GHz	0.5	1	1	2002
CPU2	Intel Pentium IV at 2.8 GHz	0.5	1	1	2003
CPU3	AMD Athlon XP3000+ at 2 GHz	0.5	1	1	2003
CPU4	AMD Sempron 2800+ at 2 GHz	0.5	1	1	2004
CPU5	AMD Athlon XP3200+ at 2.11 GHz	1	1	1	2005
CPU6	AMD Athlon XP4000+ at 2.11 GHz	0.5	2	1	2006
CPU7	AMD Phenom8450 at 2 GHz	2	3	8	2008
CPU8	Intel CI7 2600 at 3.40 GHZ	4	4	1	2011

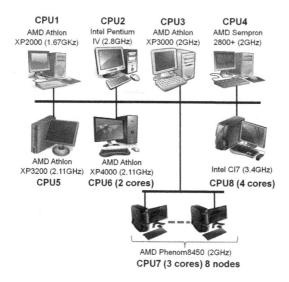

Fig. 2. Heterogeneous computing platform for experiments

processor vendors. The details about our heterogeneous environment are shown in Table 3, where specifications of each node (CPUi) are included, regarding processor, clock speed, memory, number of cores, number of nodes, and release year. From that table we can see that there is a number of commodity commercial computers and the release date of nodes corresponds to a wide range of years. Our heterogeneous computing platform is made up of one machine of each class of processor, except in the case of CPU7, where eight identical machines are included. All these machines are connected by a Gigabit Ethernet. Figure 2 sketches the heterogeneous hardware.

5.2   Ranking Using Standard Benchmark Software

We have used scientific benchmarking software to obtain a quantitative measure of the speed of the different processors involved in our heterogeneous platform.

Table 4. Normalize score value for each benchmark and CPU

	Dhrystone 1	Dhrystone 2	Whetstone	Livermore	Linpack	Mean
CPU1	4.40	4.45	2.46	3.52	3.51	**3.67**
CPU2	4.66	4.45	2.48	3.27	2.25	**3.42**
CPU3	3.53	3.69	2.29	3.15	3.16	**3.17**
CPU4	3.66	3.71	2.39	3.11	3.04	**3.18**
CPU5	3.18	3.35	2.05	2.78	2.83	**2.84**
CPU6	3.04	3.22	1.97	2.67	2.83	**2.75**
CPU7	3.90	3.25	2.00	2.57	2.83	**2.91**
CPU8	1.00	1.00	1.00	1.00	1.00	**1.00**

The reason behind the use of such benchmarking software is the difficulty to know in advance which class of program will be run in the heterogeneous platform. Particularly, a GA manages many data types: integers (population size, number of generations, alleles, etc.), floats (fitness values, probabilities, alleles, etc.), among others. The different considered data types also depend of the problem to be solved. Consequently, the corresponding compiled program is full of different types of data. These standard benchmarks are very popular and widely used in the professional computer market. So, we first go for them and then, in the second part of this phase, we will try to fine tune the findings got here.

Six different widely-used benchmarking programs have been employed, namely Whetstone [10], Dhrystone version 1 and 2 [28], Livermore Loops [21], and finally Linpack [13] (see [5] for a classification and detailed explanation). The source code can be obtained in Roy Longbottom's PC Benchmark Collection[1]. By running these benchmarking programs on each machine we obtain the scores after an operation that takes a few seconds (less than 10 s on average), independently of the benchmark used.

Table 4 presents the score values with respect to the fastest processor (baseline processor). Each benchmark gives different rankings for the processors, but all agree that CPU8 is the fastest one and CPU1 or CPU2 the slowest ones. Due to the diversity in the ranking obtained by each software benchmark, the last column of Table 4 shows the mean normalize score value for each machine. This new value is considered as the final score for each machine in our heterogeneous platform, generating the following global ranking of CPUs: 1, 2, 4, 3, 5, 7, 6, and 8. This ranking confirms our assumptions about the performance of each machine.

5.3   Ranking Using Traditional dGAs

In this section, we present an analysis of the relative velocities of the processors belonging to the heterogeneous computing environment from another point of

---

[1] http://www.roylongbottom.org.uk/.

Table 5. Relative velocities between processors discriminated by instances (fastest processor CPU8 as baseline processor). Stop condition: to reach a maximum number of generations

	KP1-1k	KP1-10k	KP2-1k	KP2-10k	KP3-1k	KP3-10k	KP4-1k	KP4-10k	Mean
CPU1	4.40	4.39	4.67	4.68	4.60	4.60	4.75	4.75	**4.61**
CPU2	3.50	3.49	3.58	3.58	3.55	3.56	3.58	3.58	**3.55**
CPU3	2.66	2.64	2.77	2.76	2.78	2.78	2.81	2.82	**2.75**
CPU4	3.20	3.19	3.39	3.39	3.50	3.51	3.60	3.59	**3.42**
CPU5	2.43	2.42	2.54	2.55	2.58	2.58	2.63	2.64	**2.55**
CPU6	2.23	2.23	2.32	2.33	2.36	2.36	2.43	2.43	**2.34**
CPU7	2.25	2.24	2.35	2.34	2.42	2.42	2.45	2.44	**2.36**
CPU8	1.00	1.00	1.00	1.00	1.00	1.00	1.00	1.00	**1.00**

view: the execution of the dGA for each problem instance. This analysis is based on the fact that a processor with a good score, running a general benchmark, may not be relevant for our metaheuristic, because of its specific kind of operations. The stop condition has been to reach a maximum number of generations (5000 for all the instances) in order to measure the elapsed time for the dGA executed by each processor of Table 3 under the same computational effort. The parameters of the dGA are the ones listed in Table 2.

Table 5 shows the relative performance of the processors regarding the baseline processor (CPU8) for each problem instance. From the analysis of the previous table, we can see that there are important differences in velocities between the considered processors. For example, CPU1 is more than four times slower than CPU8. CPU6 and CPU7 show similar relative velocities and their position in the ranking is hard to differentiate. A machine rank can be established from slow to fast processors: 1, 2, 4, 3, 5, 6/7, 8. Most of the used scientific benchmark software packages rank the processors in a different way than the running time does (let us compare Tables 4 and 5), but the rank matches the one obtained by averaging the mean scores of the standard benchmark software (last column of Table 4). Thus, we successfully matched the general benchmarks and our particular dGA benchmark into one single common ranking, what offers us a grounded way of thinking in the relative speeds of the processors so as to use it in the design of the algorithms later.

5.4   Mathematical Approximation of Running Times

As mentioned in Sect. 3, the application of the HAPA methodology requires the computation of the relative velocity factor between machines involved in the heterogeneous platform. Therefore, a comparison between benchmark and runtime results is carried out, by using the mean normalized benchmark scores and the normalized relative velocities shown in the two previous sections. However, it should be impractical to run also (all) the target algorithm to determine the

Table 6. Coefficient (left) and fitting statistics (right)

coeff.	value
a	-0.766
b	-4.508
c	1.616
d	1.074

metric	value
R^2	0.9708
Adjusted R^2	0.9489
Root means squared error	0.1698

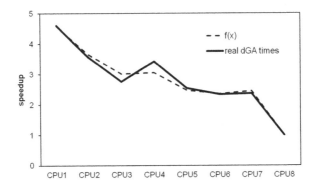

Fig. 3. Real time fitting

ranking. Therefore, we need a mathematical function to reflect that relation, as a way to predict the running times depending on the node where an island is to be assigned, and to enable us to compute the velocity factor. Also, determining this mathematical function which encapsulates the practical knowledge allows us to extend the use of the methodology to other algorithm applications.

For that purpose (encapsulating the knowledge of the previous experiments into a mathematical tool), we have used the Open Source project *Pythonequations*[2]. *Pythonequations* is a collection of Python equations that can fit themselves to both 2D and 3D data sets (curve fitting and surface fitting).

Among the best fitting equations, we have chosen the Hyperbolic G 2D equation because it was the best one in the consistency with the expected behaviour of the processors, with a reasonably low number of coefficients. The resulting formula is outlined in Eq. 2 while the coefficient values and error statistics of the regression are shown in Table 6.

$$f(x) = \frac{a \times x}{b + x} + \frac{c \times x}{d + x} \tag{2}$$

In Eq. 2, the x value corresponds to the mean scores obtained by scientific benchmark for each CPUi machine (shown in the last column of Table 4) and the $f(x)$ value means the mean relative velocities regarding execution times of the dGA for that machine. This last value is important to apply the methodology devised in Sect. 3, because it is used to compute the *velocity factor* (VF) value

---

[2] https://code.google.com/p/pyeq2/.

which is necessary to set some parameters of every local dGA, such as maximum number of generations or migration frequency.

In Fig. 3 we can see the curves representing the $f(x)$ values for each machine calculated by using Eq. 2, and the real running times obtained in our tests. Consequently, Eq. 2 can be efficiently used to predict the running times of the dGA and the $f(x)$ can be used to compute the VF for a particular machine. For example, if we should need to incorporate a new machine to the computational environment in a future, we should only obtain the scores for each software benchmark, and compute the average. This last value corresponds to the x value in Eq. 2. After that, we obtain an approximation of the mean score time regarding execution times of the dGA for that incorporated machine. At this point, it is important to remark that no new runs of the dGA is needed in this new machine. Finally, we only have to compute the VF for that machine in order to be used in the algorithm.

The considered hardware platform is a very heterogeneous cluster (see Table 3), then there is no reason to think that the method (not the concrete results) will not work for other machines, because the same measures could be applied and the same conclusions should be produced: that is the basis of a methodology.

6   HAPA Phase 2: Design Your Algorithm

This section is dedicated to the second phase of the methodology, which consists in the design of the dGA fitted to be run in the heterogeneous hardware platform. For that purpose, we describe the two possible instantiations of the proposed methodology in the design of a dGA.

6.1   DGA_HAPA: A First Algorithm Derived from HAPA

In this first algorithm derived from the HAPA methodology, we search for the unification of the completion time of the algorithm on all the islands. For that, the derived dGA establishes in an online way how many generations a subpopulation has to evolve depending on its underlying CPU. If the population in the fastest processor evolves for G number of generations (max_{gens} parameter), each island i of our dGA_HAPA executing in a slower processor B has to set its local number of generations (max_{gens-i}) to a value equal to G divided by $VF(B)$ (the computed relative velocity factor for processor B). In a first step, each island of our dGA derived from the HAPA methodology asks for the features of the processor where it was launched (obtaining the $VF(B)$ value), and also reads the configuration file where the number of generations parameter is generically set to G. With this information it proceeds to locally compute the maximum number of generations. Algorithm 1 shows how a dGA$_i$ node is developed from the HAPA methodology.

The rationale in this design is to avoid big deviations during the search process, usually making just one or two of the islands to produce interesting

---

Algorithm 1. dGA$_i$ using the HAPA methodology

VF=get(local Velocity Factor of processor)
$max_{gens-i} = max_{gens}/VF$
$t = 0$; {current generation}
initialize($P_i(t)$);
evaluate($P_i(t)$);
while ($t < max_{gens-i}$) **do**
    $P_i'(t) = $ evolve($P_i(t)$); {recombination and mutation}
    evaluate ($P_i'(t)$);
    $P_i'(t) = $ send/receive individuals from dGA_j;
    $P_i(t + 1) = $ select new population from $P_i'(t) \cup P_i(t)$;
    $t = t + 1$;
end while

---

results, while the rest are stuck in old and lower quality solutions. Heterogeneity could easily produce this behavior, and we want to find out whether our two research questions (accuracy and efficiency) hold in this HAPA design.

6.2 DGA_HAPA-FM: A Second dGA Derived from HAPA

The proposed HAPA methodology can, of course, be used to derive other algorithms. As a second example, we here now considered the migration frequency in such a way that all islands will finally receive the same number of migrants during the evolution. The aim is to prevent that fast islands probably end their evolution without information from slow islands, also a normal non-desired behavior of plain heterogeneous algorithms.

The same process as the one previously described to compute the maximum number of generations on each island depending on the processor's velocity is carried out (similar pseudocode, thus not shown). If the migration frequency is set to M generations on the fastest island, then this value is divided by the relative velocity factor $VF(B)$ of the rest of processors executing other island GAs. With this operation, the number of generations between consecutive steps of sending/receiving (the migration frequency parameter) is obtained, thus achieving an indirect numerical synchronization between the islands. This implicit synchronization comes from the proportional rate that the islands have between consecutive sending (in that way all islands exchange the same number of individuals during the evolution on average).

7   HAPA Phase 3: Obtaining the Results

This section is devoted to the third phase of the methodology, where the HAPA performance in the developing of dGAs to be run in the heterogeneous hardware platform is evaluated. For that purpose, we consider the two dGAs described in the previous section: dGA_HAPA and dGA_HAPA-FM. The heterogenous environment considered for their execution is made up of one node of each CPUi, with $i \in \{1, \ldots, 8\}$.

For comparison purposes we include the results of traditional dGAs under different homogeneous execution scenarios: (i) running a traditional dGA in a concurrent manner, i.e., mapping the eight islands onto only one processor, denominated dGA_1CPUi (CPUi with $i = 1, 2, \ldots, 8$) thus obtaining a dGA variant for each CPU shown in Table 3 and, (ii) running a traditional dGA in parallel (dGA_hom) using a parallel homogeneous configuration, where each island is mapped to a ring of eight processors (using the CPU7 hardware configuration). With this comparison we aim at determining whether using HAPA in dGAs has been useful to provide similar and even better results than dGAs running in homogeneous hardware platforms. The stop condition for all algorithms has been established to reach the optimum solution for each instance or the maximum number of generations, whichever happens first.

In what follows, we measure basic parameters such as the hit rate, numerical effort/time to locate a solution, and speedup. The goal is to offer a thorough study of dGAs using HAPA as a way to validating this idea when executing in heterogeneous computers. The section ends with an analysis of the search diversity to examine the behavior of the studied dGAs.

7.1 Hit Rate Analysis

The first considered quality indicator will be the hit rate, i.e., the number of times at which an algorithm finds the optimal solution for an instance, out of a constant number of 30 independent runs. Table 7 displays the obtained results. Let us begin by analyzing the results of the dGA running on one single processor (dGA_1CPUi) for all our processors independently. As expected, the different dGA_1CPUi present similar hit rate values, which can be explained by the fact that the dGA is exactly the same in all the tests (numerically speaking), being the processor velocity the only difference between them. The dGA running in the homogeneous hardware (dGA_hom), where each island is mapped onto a processor (8 islands and 8 equal processors), also obtains similar results as the different dGA_1CPUi, except for KP3-10k whose hit rate is nearly the half of the ones of dGA_1CPUi. Finally, the dGAs executed in a heterogeneous computing environment (dGA_HAPA and dGA_HAPA-FM) are able to find the optimal solution in a similar number of runs as those of the different dGA_1CPUi and dGA_hom. In general, the hit rate values are higher than the 60 % for the three first instances (except for KP2-1k). In the case of instances KP2-10k, KP4-1k, and KP4-10k the dGA was in general not able to find the optimal solution independently of the hardware used, but our actual target is not problem solving, but algorithm design.

7.2 Numerical Effort Analysis

Table 8 shows the numerical effort to locate a solution, i.e., the number of evaluations of the objective function needed to locate the optimum (Table 1). There are

Table 7. Hit rate obtained by dGA_1CPUi, dGA_hom and HAPA variants for each problem size

	dGA_1CPUi								dGA_hom	dGA_HAPA	dGA_HAPA-FM
	1	2	3	4	5	6	7	8			
KP1-1k	93	83	77	87	83	73	93	80	87	63	80
KP1-10k	97	90	77	93	90	100	100	93	90	93	93
KP2-1k	57	77	70	87	73	73	77	70	70	60	50
KP2-10k	0	0	0	0	0	0	3	0	0	0	0
KP3-1k	13	13	7	0	3	7	7	7	7	7	13
KP3-10k	47	40	27	40	43	50	43	33	27	17	27
KP4-1k	0	3	3	0	3	0	0	3	0	0	3
KP4-10k	0	0	0	0	0	7	0	0	3	0	0

Table 8. Numerical effort obtained by dGA_1CPUi, dGA_hom and dGA_HAPA variants for solving each problem size

	dGA_1CPUi								dGA_hom	dGA_HAPA	dGA_HAPA-FM
	1	2	3	4	5	6	7	8			
KP1-1k	65459	67646	64835	74102	69062	59774	54932	55964	72635	47739	**35991**
KP1-10k	56188	60197	46273	91385	50131	55569	55569	56493	49282	69662	**42794**
KP2-1k	88733	91411	92270	126219	92788	**67321**	92492	88828	94397	81542	83723
KP2-10k	–	–	–	–	–	–	110439	–	–	–	–
KP3-1k	145552	145599	144567	–	202702	157149	135002	220203	190293	**63336**	67106
KP3-10k	157951	151096	173798	164674	150141	157023	176838	146770	154821	**86918**	99457
KP4-1k	–	106168	93726	–	127657	–	–	252246	–	–	**61282**
KP4-10k	–	–	–	–	–	154351	–	–	110824	–	–

no important differences between the results of the dGA_1CPUi and dGA_hom. However, an encouraging finding is this: dGA_HAPA variants reach the optimal solutions in a smaller number of evaluations than the rest, except for KP2-1k. The Kruskal-Wallis test confirms this situation, indicating that there are statistically significant differences between groups of ranks (p-values lower than 10^{-5}). There seems to be no statistical difference between the dGA executed in a uniprocessor environment. However, dGA_HAPA-FM presents statistically significant differences with respect to the rest of the algorithms (in its favor) for KP1-1k, KP1-10k and KP4-1k. In the case of KP2-10k and KP4-10k the optimal solution was too difficult to find for all the algorithms.

In the first two instances, the percentage of numerical effort reduction achieved by dGA_HAPA-FM with respect to dGA_1CPUi and dGA_hom is between 22 % (dGA_1CPU3 for KP1-10k) and 61 % (dGA_1CPU4 for KP1-10k) and for KP2-1k the values decrease by more than the 50 %. For the rest of the instances the values decrease by the 75 %. Analyzing the number of evaluations of dGA_HAPA with respect to dGA_1CPUi and dGA_hom for KP2-1k, a reduction of over 30 % is achieved, while for KP3-1k and KP3-10k the values decrease by more than the 67 %. Consequently, the exploration of the search space made by dGA_HAPA and dGA_HAPA-FM is not the same as for the other algorithms, because the final number of function evaluations is fairly smaller. The truly interesting point is that, even with such a lower effort, the hit rate is highly competitive compared to the ones of the other non-HAPA algorithms analyzed.

All the previous observations, i.e., similarities in hit rate values and numerical effort, represent a promising result, remarking the feasibility of merging different old hardware with new machines as a unified heterogeneous platform to execute a dGA. With these results, we can positively answer the first question (Q1) made in the introduction (Sect. 1): HAPA can help in designing new algorithms for heterogeneous hardware with competitive times (compared to new hardware) and with better problem-solution accuracy.

7.3 Runtime Analysis

Another interesting measure for a dGA is the total runtime needed to reach a solution, which is shown in Fig. 4. In this case, the mean runtime in seconds is displayed. From the point of view of the uniprocessor configurations, we can see that dGA_1CPU8 is the fastest configuration, followed by dGA_1CPU6, dGA_1CPU7, and so on, i.e., the resulting ranking matches the underlying technologies' state of the art, an expected situation. The two HAPA variants are the fastest hardware configurations (lowest execution time). All the differences are statistically significant according to the Kruskal-Wallis test (p-values bellow to 10^{-5}), dGA_HAPA-FM and dGA_hom have ranks significantly different from dGA_HAPA. These results are the basis to answer the second question (Q2) formulated in Sect. 1: we do not observe a degradation in the execution time with the use of the heterogeneous computing environment (slower computers on it), which again is an interesting finding. Homogeneous dGAs can be only better, in theory, if all CPUs involved in the computing environment are equal to the fastest hardware configuration belonging to the heterogeneous parallel platform.

Related to the runtime measure, we have also studied the speedup s_m, which compares the run time of the parallel algorithm on one processor against the run time of the same algorithm on m processors to solve a particular problem [1]. For non-deterministic algorithms, the speedup is the ratio between the mean execution time of the dGAs on a uniprocessor configuration, denoted as $E[T_1]$, and the mean execution time of the dGAs on m processors, denoted as $E[T_m]$ with $m = 8$ (see Eq. 3). The speedup has been computed here with respect to the fastest processor (CPU8, worst case analysis).

$$s_m = \frac{E[T_1]}{E[T_m]} \tag{3}$$

Figure 5 graphically shows the speedup values for each dGA (the line in this figure expresses the ideal linear speedup). The dGA_HAPA-FM speedup is the best of the three algorithms for the most of the instances. Although the speedup is sub-linear ($s_m < m$), the heterogeneous results are quite good because they are

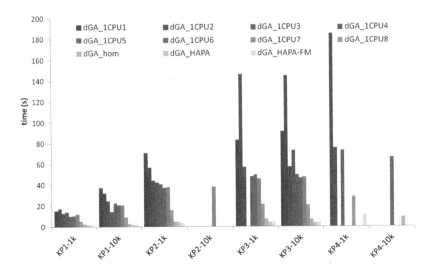

Fig. 4. Execution times obtained by the dGA for each problem size

approximately at 80 % of the ideal speedup value, except for instances KP1-1k, KP4-1k and KP4-10k where the values are very small (less than 2.7). This last fact could be explained by the huge difference between the power of the different hardware configurations used: we must remember that the reference point for speedup is the best performing processor, a very fast one in global terms.

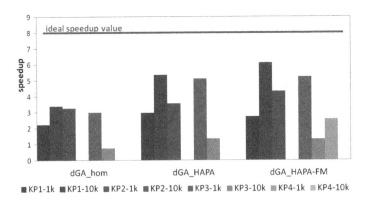

Fig. 5. Speedup obtained by the dGAs for each problem size

7.4   Search Diversity Analysis

We proceed now with an analysis of the evolution of the fitness and the diversity of the dGA_hom, dGA_HAPA and dGA_HAPA-FM. For this purpose, we are

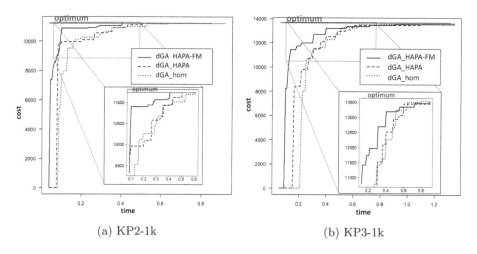

(a) KP2-1k   (b) KP3-1k

Fig. 6. Evolution of the mean fitness for two illustrative problems

going to track the value of the best fitness and the mean population entropy along the search, which are the mean of ten runs made for each algorithm. We did just ten runs because it is a fairly stable process, as the results will show.

Figure 6 shows the evolution of the best fitness along the execution of each algorithm for some problem instances used as example (similar situations are observed in the rest of the instances). In the bottom right corner of each subfigure, we make a zoom into detailed moments of the evolution. Regardless of the problem dimension, the population diversity of dGA_HAPA-FM leads to good solutions faster than the rest. This observation suggests that the cooperation of slow processors and faster processors, together with the adjustment of the migration frequency, lead to work out high quality solutions. After an initial period of evolution, the curves belonging to dGA_HAPA and dGA_hom overlapped to the dGA_HAPA-FM.

Figure 7 presents the evolution of the mean population entropy along the search for each algorithm analyzed (measured for every bit position in all the individuals of a population). In this case, we show the accumulated entropy values over time. We are going to analyze this figure and compare values with the aforementioned evolution of the fitness. We can see how the fitness curve has a maintained growth rate for the dGA_HAPA-FM population, suggesting that it was able to sustain a higher diversity within its population during the whole search, which helps to produce good solutions. The algorithm dGA_hom reaches good entropy levels compared to the dGA_HAPA one.

Another factor to analyse is the quality of immigrants with respect to the best solution in the target population. Our aim is to give a measure of the degree of cooperation between the different subpopulations of an algorithm. We compute the percentage of times the incoming solution has a higher quality than the

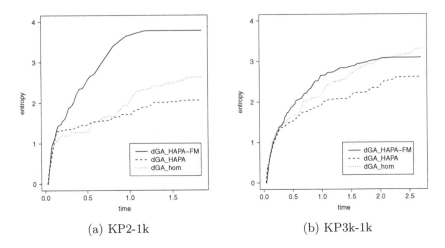

(a) KP2-1k   (b) KP3k-1k

Fig. 7. Evolution of the entropy for two illustrative problems

Table 9. Percentage of acceptance of the received solutions into de local subpopulation

%	dGA_hom	dGA_HAPA	dGA_HAPA-FM
KP1-1k	94.48	80.22	90.03
KP1-10k	91.15	71.90	84.90
KP2-1k	86.59	76.56	79.26
KP2-10k	82.01	78.54	73.43
KP3-1k	83.56	71.07	72.98
KP3-10k	82.19	73.63	73.95
KP4-1k	76.31	70.25	70.11
KP4-10k	75.54	67.77	62.20
Mean	83.60	73.74	75.86

best solution in the target population. Table 9 shows the mean percentages for all dGAs and problem instances. In general, the percentages decrease as the complexity of the instances increases. The dGA_hom algorithm obtains the higher percentages than the rest, a typical excess of elitism that renders unproductive the algorithm. In most of the cases, the dGA_HAPA-FM has a higher percentage of acceptance than the dGA_HAPA, meaning that the received solution is incorporated to the target population. In a way, previous observations indicate the beneficial influence of collaboration between the subpopulations in different stages of the evolution. The exchange of solutions not only benefits slower islands (by the reception of optimized solutions) but also benefits faster islands because of solutions having portions of the optimum in them.

8   Conclusions and Future Research

This article deals with the execution of a dGA using heterogeneous computing resources, where the processing nodes show a high level of heterogeneity: different CPUs belonging to a wide range of fabrication years and technologies. We developed a methodology, called HAPA, to deal with that heterogeneity and the execution of metaheuristics. HAPA consists of three phases: (i) the computation of a ranking of processors in order to know the platform (an offline phase), (ii) the algorithm design derived from the previous phase, and (iii) the validation of the proposed dGAs (an online phase). In this work, the HAPA methodology was applied to regulate the stopping conditions (dGA_HAPA algorithm) and the migration frequency (dGA_HAPA-MF algorithm).

We have performed a set of tests in order to assess the performance of our proposal, and we compared both algorithms derived from the HAPA methodology against a dGA running in a homogeneous computing environment. The results indicate that similar levels of accuracy and efficiency can be attained, but with a lower number of function evaluations, by using the proposed HAPA methodology, thus confirming RQ1. We have shown that the dGA_HAPA-MF algorithm can perform a search in a faster way than the homogeneous dGAs, while maintaining a higher diversity within the population, exhibiting a better balance between exploration and exploitation. Therefore, we can also finally confirm RQ2.

In short, we have contributed in this article with a way of avoiding pure ad-hoc design of algorithms for heterogeneous platforms, as well as we give a new line of research in how to inject hardware knowledge into software parameters of the algorithms. In addition of being innovative, this has shown to also be numerically competitive and within reduced run times.

Further research is necessary to understand the effects of factors such as the influence in the topology of communication between islands. We will also consider a self-control of the parameters of the distributed genetic algorithm during the evolution under a heterogeneous environment taking care of both, numerical and hardware issues.

Acknowledgments. We acknowledge the UNLPam, the ANPCYT, CONICET and PICTO-UNLPam-0278 in Argentina from which Dr. Salto receives regular support. The work of Prof. Alba has been partially funded by the University of Málaga UMA/FEDER FC14-TIC36, programa de fortalecimiento de las capacidades de I+D+I en las universidades 2014–2015, de la Consejería y Economía, Innovación, Ciencia y Empleo, with European FEDER, and also by the UMA Project 8.06/5.47.4142 with the VSB-Technical University of Ostrava (CR). Finally, we acknowledge the funding by the Spanish MINECO project TIN2014-57341-R (http://moveon.lcc.uma.es).

References

1. Alba, E.: Parallel evolutionary algorithms can achieve super-linear performance. Inf. Process. Lett. **82**(1), 7–13 (2002). Elsevier

2. Alba, E.: Parallel Metaheuristics: A New Class of Algorithms. Wiley, New Jersey (2005)
3. Alba, E., Nebro, A.J., Troya, J.M.: Heterogeneous computing and parallel genetic algorithms. J. Parallel Distrib. Comput. **62**, 1362–1385 (2002)
4. Alba, E., Troya, J.M.: Analyzing synchronous and asynchronous parallel distributed genetic algorithms. Future Gener. Comput. Syst. **17**(4), 451–465 (2001)
5. Domínguez, J., Alba, E.: A methodology for comparing the execution time of metaheuristics running on different hardware. In: Hao, J.-K., Middendorf, M. (eds.) EvoCOP 2012. LNCS, vol. 7245, pp. 1–12. Springer, Heidelberg (2012)
6. Baugh, J., Kumar, S.: Asynchronous genetic algorithms for heterogeneous networks using coarse-grained dataflow. In: Cantú-Paz, E., et al. (eds.) GECCO 2003. LNCS, vol. 2723, pp. 730–741. Springer, Heidelberg (2003)
7. Bazterra, V.E., Cuma, M., Ferraro, M.B., Facelli, J.C.: A general framework to understand parallel performance in heterogeneous clusters: analysis of a new adaptive parallel genetic algorithm. J. Parallel Distrib. Comput. **65**(1), 48–57 (2005)
8. Branke, J., Kamper, A., Schmeck, H.: Distribution of evolutionary algorithms in heterogeneous networks. In: Deb, K., Tari, Z. (eds.) GECCO 2004. LNCS, vol. 3102, pp. 923–934. Springer, Heidelberg (2004)
9. Chong, F.: Java based distributed genetic programming on the internet. Technical report, School of Computer Science, University of Birmingham (1999)
10. Curnow, H.J., Wichmann, B.A.: A synthetic benchmark. Comput. J. **19**(1), 43–49 (1976)
11. Dominguez, J., Alba, E.: Ethane: A heterogeneous parallel search algorithm for heterogeneous platforms. In: DECIE 2011 (2011)
12. Dominguez, J., Alba, E.: Dealing with hardware heterogeneity: a new parallel search model. Nat. Comput. **12**(2), 179–193 (2013)
13. Dongarra, J.: Performance of various computers using standard linear equations software in a fortran environment. Simulation **49**(2), 51–62 (1987)
14. Alba, E., et al.: MALLBA: a library of skeletons for combinatorial optimisation. In: Monien, B., Feldmann, R.L. (eds.) Euro-Par 2002. LNCS, vol. 2400, pp. 927–932. Springer, Heidelberg (2002)
15. García-Arenas, M., Merelo, J., Castillo, P., Laredo, J., Romero, G., Mora, A.: Using free cloud storage services for distributed evolutionary algorithms. In: Proceedings of the 13th Annual Conference on Genetic and Evolutionary Computation, GECCO 2011, pp. 1603–1610. ACM, New York (2011)
16. Garey, M.R., Johnson, D.S.: COmputers and Intractability: a Guide to the Theory of NP-Completeness. Freeman, New York (1979)
17. Gong, Y., Nakamura, M., Tamaki, S.: Parallel genetic algorithms on line topology of heterogeneous computing resources. In: Proceedings of the 2005 Conference on Genetic and Evolutionary Computation, GECCO 2005, pp. 1447–1454 (2005)
18. Lim, D., Ong, Y.-S., Jin, Y., Sendhoff, B., Lee, B.-S.: Efficient hierarchical parallel genetic algorithms using grid computing. Future Gener. Comput. Syst. **23**(4), 658–670 (2007)
19. Liu, P., Lau, F., Lewis, M.J., Wang, C.: A new asynchronous parallel evolutionary algorithm for function optimization. In: Merelo Guervós, J.J., Adamidis, P.A., Beyer, H.-G., Fernández-Villacañas, J.-L., Schwefel, H.-P. (eds.) PPSN 2002. LNCS, vol. 2439, pp. 401–410. Springer, Heidelberg (2002)
20. Luque, G., Alba, E. (eds.): Parallel Genetic Algorithms: Theory and Real World Applications. SCI, vol. 367. Springer, Heidelberg (2011)
21. McMahon, F.H.: The Livermore Fortran Kernels: A Computer Test of the Numerical Performance Range. Lawrence Livermore National Laboratory (1986)

22. Meri, K., Arenas, M., Mora, A., Merelo, J., Castillo, P., Garca-Snchez, P., Laredo, J.: Cloud-based evolutionary algorithms: An algorithmic study. Natural Comput. **12**(2), 135–147 (2013)
23. Mostaghim, S., Branke, J., Lewis, A., Schmeck, H.: Parallel multi-objective optimization using master-slave model on heterogeneous resources. In: IEEE Congress on Evolutionary Computation, CEC 2008. (IEEE World Congress on Computational Intelligence), pp. 1981–1987 (2008)
24. Pisinger, D.: A minimal algorithm for the 0–1 knapsack problem. Oper. Res. **45**, 758–767 (1997)
25. Pisinger, D.: Core problems in knapsack algorithms. Oper. Res. **47**, 570–575 (1999)
26. Pisinger, D.: Where are the hard knapsack problems? Comput. Oper. Res. **32**, 2271–2282 (2005)
27. Tanese, R.: Distributed genetic algorithms. In: Proceedings of the Third International Conference on Genetic Algorithms, pp. 434–439 (1989)
28. Weicker, R.P.: Dhrystone: a synthetic systems programming benchmark. Commun. ACM **27**(10), 1013–1030 (1984)

Eroca: A Framework for Efficiently Recovering Outsourced Ciphertexts for Autonomous Vehicles

Xu An Wang[1,3]([✉]), Jianfeng Ma[2], Yinbin Miao[1], and Kai Zhang[1]

[1] School of Telecommunications Engineering,
Xidian University, Xi'an, People's Republic of China
wangxazjd@163.com
[2] School of Cyber Engineering, Xidian University, Xi'an, People's Republic of China
[3] Engineering University of Chinese Armed Police Force,
Xi'an, People's Republic of China

Abstract. In the near future, the next generation (5G) telecommunication network with high speed will become a reality. Autonomous vehicle system without drivers are a typical application of 5G network, for it can connect base stations, autonomous vehicles and computing centers such as traffic information clouds in a very flexible, truly mobile and powerful way. To ensure the security and privacy of autonomous vehicle system, a promise way is to encrypt the real time traffic information and upload the ciphertexts to the center cloud for easily sharing road traffic information among the vehicles. To share these real time traffic information without sacrificing privacy, attribute based encryption (ABE) and block ciphers like AES are promising tools for encrypting these large traffic information. But a basic fact for the autonomous vehicle system is that, the vehicles need to continuously update the traffic information for really catching the road's real time traffic status, and these updates are often little compared with the status a moment ago. In this paper, we consider the problem of how to retrieve and update the data from the early encrypted file in the cloud efficiently. We propose the notion of attribute based encryption with sender recoverable (ABE-SR). Compared with ABE, ABE-SR can easily achieve *message recoverable and updatable for the encrypter*. We give a concrete ABE-SR scheme, discuss its features compared with the traditional ABE and prove its security. Based on ABE-SR, we propose a new framework for efficiently recovering outsourced ciphertexts for autonomous vehicles: Eroca. Finally, we give the roughly evaluation results, which show our proposal framework is practical.

Keywords: Outsourced ciphertexts · Attribute based encryption · Efficiently recovering · Autonomous vehicles · 5G network

© Springer-Verlag Berlin Heidelberg 2015
N.T. Nguyen et al. (Eds.): Transactions on CCI XIX, LNCS 9380, pp. 126–139, 2015.
DOI: 10.1007/978-3-662-49017-4_8

1 Introduction

1.1 5G Telecommunication Network and Autonomous Vehicle System

Along with the rapid development of 5G technologies recently, many corporations like Huawei show great interest on switching from 4G to this new telecommunication network. In the near future around 2020, the 5G network will probably become a reality. The service of 5G will be infinitely richer, more complex and better quality of experience. 4G has a key feature of supporting high data rate (up to 1 Gbit/s) on the downlink. However, 5G will have a super high speed around 10 Gbit/s, which is around 10–100 times of 4G. Besides the super high speed, 5G will also pay attention on coverage and user experience. For example, "low latency" is one of the most significant feature of 5G, it can support billions of embedded equipments, high density of communication interaction, and high speed mobile systems simultaneously. It can change our experience on using telecommunication with 3G or 4G. Many wonderful things which were unimaginable can be possible when 5G network is available.

Autonomous vehicle system is one of the promising typical applications with 5G. Autonomous vehicle system will revolutionize our exist traffic management paradigm. All the mobile cars are controlled by computers or embedded intelligent equipments, and the city needs an intelligent center like a traffic cloud to properly dispatch the real time traffic dynamically. In this system, the mobile cars need interact with the cloud center and the near cars very often. The current 4G technique can not support the so high dense interaction without almost no latency, but a little latency with 0.5 s will be enough to cause a disaster when the mobile car is running with high speed. But 5G can support these high dense interaction with almost no latency, which is crucial for autonomous vehicle system. Furthermore, 5G allows directly dense communication among mobile cars like D2D communication, which is another crucial factor for smoothly running autonomous vehicle system.

1.2 Attribute Based Encryption for Sharing Real-Time Traffic Information

However when the traffic information switch frequently in the whole city's traffic management system, the security and privacy of the mobile car users can be broken more easily. Thus we need to consider implement security mechanisms for autonomous vehicle system from the beginning. A common method to ensure security and privacy is encrypting the sensitive traffic information before uploading them to the traffic cloud.

Attribute based encryption (ABE) [1–5,9] is a new one-to-many encryption paradigm which is very suitable for fine-grained control on accessing the contents of the ciphertexts, thus it is also suitable for sharing the sensitive traffic information for autonomous vehicle system. For running autonomous vehicle system, it is crucial for more and more autonomous vehicles upload their personal traffic data in the public cloud for the reasons of saving the storage, while at the same time achieving easily access at anywhere, cheaply management etc. properties as well. Furthermore, they often need to share the content with other vehicles, base stations, managers etc., but prohibit to let the personal data be completely public.

1.3 Motivation and Contribution

Many proposals on how to use ABE in cloud data storage system have been put forward [6–8,10–12]. However, they often omit a basic fact: the data owner always need to retrieve and update the data from the early encrypted file in the cloud. All the ABE schemes until now can not handle this problem easily, they do not have the ability of recovering the contents from the upload ciphertexts once the data owner have encrypted the contents under the prescribed attributes or policies.

Autonomous vehicles need to share the data with closely vehicles, vehicle base stations according to different polices, but they are often with limited resources such as little memory, a limited storage space, weak computation capability etc. they also do not maintain the traffic content copy in the local embedded equipments. But later they have to retrieve the upload encrypted file, for editing the traffic information. But traditional ABE can not satisfy this requirement. Thus we need to consider how to recover and update the data from the early encrypted file. Autonomous vehicles (Data owner) can update their own data easily is an obvious basic requirement for smoothly running the system.

We illustrate this situation in the following Fig. 1. Vehicle V1 encrypts the real traffic information under the attributes ($Attribute1$: $StreetB$) and ($Attribute2$: $Speed > 70\,km/h$), and send them to the cloud, Vehicle V2 can decrypt and share the traffic information because its secret key corresponding to the access structure ($Attribute1$: $StreetB)AND(Attribute2$: $Speed \geq 70\,km/h$), but Vehicle V1 itself can not decrypt the ciphertext C once it has outsourced it to the cloud.

To solve this problem, we introduce a new variant of ABE: ABE-SR, which is almost the same as the traditional ABE except *the encrypter* can recover the plaintexts from the ciphertexts. The main idea is letting the data owner embedding a trapdoor (a fixed randomness) when generating the ciphertexts, which will be later used to recover the data contents. Compared with ABE, ABE-SR can easily achieve *easily message recoverable and updateable for the encrypter*. These properties are very desirable in practical applications like cloud storage. Thus it can be used to efficient update the outsourced ciphertexts for the vehicles, especially these updates are quite often due to the dynamically changed traffic status. The notion of ABE-SR can be seen as an extension of Wei et al.'s notion of public key encryption with sender recoverable [13,14], which also aimed at how to recover the contents from the ciphertexts for the encrypter. We give a concrete KP-ABE-SR scheme, discuss its features compared with the traditional ABE and roughly analysis its security.

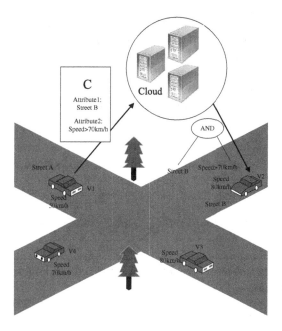

Fig. 1. ABE for sharing real-time traffic information

1.4 Organization

In Sect. 2, we give the definition and security model of key-policy ABE-SR, construct such a scheme and analysis its security. In Sect. 3, we give the system model, describe the security objectives, and show why the ABE-SR schemes can satisfy these objectives. Finally, we discuss how to extend the idea of key-policy ABE-SR (KP-ABE-SR) to ciphertext-policy ABE-SR (CP-ABE-SR) and conclude our paper.

2 Definition and Security Model for KP-ABE-SR

A Key-Policy Attribute Based Encryption with Sender Recover scheme consists of four algorithms:

1. Setup. This randomized algorithm takes the security parameter as input, and outputs the public parameters PK, the master key MK.
2. Encryption. This randomized algorithm takes a message m, a set of attributes γ, a fixed randomness R, and the public parameters PK as input, and outputs the ciphertext E.
3. Key Generation. This randomized algorithm takes the access structure $\mathbb{A}$, the master key MK and the public parameters PK as input, and outputs a decryption key D.

4. Decryption. This algorithm takes the ciphertext E that was encrypted under the set γ of attributes, the decryption key D for access control structure $\mathbb{A}$ and the public parameters PK as input, and outputs the message M if $\gamma \in A$ or $\perp$ which indicates the ciphertext is not valid.
5. Recover. This algorithm takes the ciphertext E that was encrypted under the set γ of attributes by using the fixed randomness R, the public parameters PK, the fixed randomness R as input, and outputs the message M if $\gamma \in A$ or $\perp$.

We now discuss the security of an KP-ABE-SR scheme. We define a selective-set model for proving the security of the attribute based encryption with sender recover under chosen plaintext attack.

1. Init. The adversary declares the set of attributes γ that he wishes to be challenged upon.
2. Setup. The challenger runs the Setup algorithm of ABE and gives the public parameters to the adversary, he also choose a fixed randomness as the targeted sender's recovering key.
3. Phase 1. The adversary is allowed to issue queries for private keys for many access structures Aa_j, where $\gamma \notin \mathbb{A}_j$ for all j. It is also allowed to issue Recover queries on ciphertexts C for the targeted sender and get the recovering results.
4. Challenge. The adversary submits two equal length messages M_0 and M_1. The challenger flips a random coin b, and encrypts M_b with γ. The ciphertext C^* is passed to the adversary.
5. Phase 2. Phase 1 is repeated except the ciphertexts C^* can not be queried to the recovering oracle.
6. Guess. The adversary outputs a guess b' of b.

The advantage of an adversary $\mathcal{A}$ in this game is defined as $Pr[b = b'] - 1/2$.

We note that the model can easily be extended to handle chosen-ciphertext attacks by allowing for decryption queries in Phase 1 and Phase 2.

Definition 1. *An key policy attribute-based encryption scheme with sender recover (KP-ABE-SR) is secure in the selective-set model of security if all polynomial time adversaries have at most a negligible advantage in the selective-set game.*

3 A Concrete KP-ABE-RS Scheme and Its Security Analysis

3.1 Access Tree

Ciphertexts are labeled with sets of attributes and private keys are associated with access structures that control which ciphertexts a user is able to decrypt.

Access tree Γ. We denote Γ as a tree representing an access structure. Each non-leaf node of the tree which described by its children and a threshold value, represents a threshold gate. If num_x is the number of children of a node x, k_x is its threshold value, then $0 \le k \le num_x$. The threshold gate is an OR gate when $k_x = 1$, and it is an AND gate when $k_x = num_x$. Each leaf node x of the tree is described by an attribute and a threshold value $k_x = 1$. Denote $parent(x)$ as the parent of the node x in the tree. If and only if x is a leaf node, the function $att(x)$ is defined and denotes the attribute associated with the leaf node x in the tree. An ordering between the children of every node are also defined by the access tree T, that is, the children of a node are numbered from 1 to num. The function $index(x)$ returns such a number associated with the node x, where the index values are uniquely assigned to nodes in the access structure for a given key in an arbitrary manner.

Satisfying an access tree. Denote Γ as an access tree with root r. We define Γ_x as the subtree of Γ rooted at the node x. Hence Γ is the same as Γ_r. We denote it as $\Gamma_x(\gamma) = 1$ if a set of attributes γ satisfies the access tree Γ_x. We compute $\Gamma_x(\gamma)$ recursively as follows. If x is a non-leaf node, evaluate $\Gamma'_x(\gamma)$ for all children x' of node x, if and only if at least k_x children return 1, $\Gamma_x(\gamma)$ returns 1. If x is a leaf node, then if and only if $att(x) \in \gamma$, $\Gamma_x(\gamma)$ returns 1.

3.2 The Concrete KP-ABE-RS Scheme

Following is our concrete proposal, it is almost the same as [4] except the encrypter has a fixed randomness which can also be used to decrypt the ciphertext.

1. **Setup:** Let $(\mathbb{G}_1, \mathbb{G}_2)$ be bilinear groups of prime order p, and let g be a generator of $\mathbb{G}_1$. In addition, let $e : \mathbb{G}_1 \times \mathbb{G}_1 \to \mathbb{G}_2$ denote the bilinear map. We also define the Lagrange coefficient $\Delta_{i,S}$ for $i \in Z_p$ and a set, S, of elements in Z_p: $\Delta_{i,S(x)} = \Pi_{j \in S, j \neq i} \frac{x-j}{i-j}$. Define the universe of attributes $U = \{1, 2, \cdots, n\}$.
2. **Key Generation**($Randomness, \Gamma$): The Private Key Generator (PKG) does the following:
   - For each attribute $i \in U$, choose random number t_i uniformly from Z_p.
   - Choose random number s uniformly in Z_p.
   - It publishes the public parameters PK

$$T_1 = g^{t_1}, \cdots, T_{|U|} = g^{t_{|U|}}, Y = e(g, g)^y$$

And the master secret key MK is:

$$(t_1, \cdots, t_{|U|}, y)$$

   - Then the PKG generates a key that allows the decrypter to decrypt a message encrypted under a set of attributes γ, if $\Gamma(\gamma) = 1$ holds. The algorithm proceeds as follows.
   (a) Choose a polynomial q_x for each node x (including the leaves) in the tree T. Starting from the root node r, these polynomials are chosen in the following way in a top-down manner. For each node x in the tree, set the degree d_x of the polynomial q_x to be one less than the threshold

value k_x of that node, that is, $d_x = k_x - 1$. Now for the root node r, set $q_r(0) = y$ and d_r other points of the polynomial q_r randomly to define it completely. Set $q_x(0) = q_{parent(x)}(index(x))$ for any other node x, and choose d_x other points randomly to completely define q_x.

(b) For each leaf node x, we give the following secret value to the user, once the polynomials have been decided:

$$D_x = g^{\frac{q_x(0)}{t_i}}$$

where $i = att(x)$.

3. Encryption(M, γ, PK, r): To encrypt a message $M \in \mathbb{G}_2$ under a set of attributes γ, the Sender choose a fixed number $r \in Z_p$ which will be used as his Recovering Key (RK), and a random number $s \in Z_p$ and publish the ciphertext as:

$$E = (\gamma, E' = MY^{rs} = Me(g,g)^{rsy}, E'' = Y^s = e(g,g)^{sy}, \{E_i = T_i^{sr}(i \in \gamma)\})$$

4. Decryption(E, D): We specify our decryption procedure as a recursive algorithm.

- We first define a recursive algorithm $DecryptNode(E, D, x)$ that takes as input the ciphertext $E = (E, D, x)$, the private key D (we assume the access tree Γ is embedded in the private key), and a node x in the tree. It outputs a group element of $\mathbb{G}_2$ or $\perp$.
- Let $i = att(x)$, if the node x is a leaf node then:
  • If $i \in \gamma$,

$$Decrypt(E, D, x) = e(D_x, E_i) = e(g^{\frac{q_x(0)}{t_i}}, g^{st_i r}) = e(g,g)^{syr}$$

• Otherwise, return $\perp$.

We now consider the recursive case when x is a non-leaf node. The algorithm $DecryptNode(E, D, x)$ then proceeds as follows: For all nodes z that are children of x, it calls $DecryptNode(E, D, z)$ and stores the output as F_z. Let S_x be an arbitrary k_x-sized set of child nodes z such that $F_z \neq \perp$. If no such set exists then the node was not satisfied and the function returns $\perp$. Otherwise, we compute

$$F_x = \prod_{z \in S_x} F_z^{\delta_{i, S'_x}(0)}$$

$$= \prod_{z \in S_x} (e(g,g)^{srq_z(0)})^{\delta_{i, S'_x}(0)}$$

$$= \prod_{z \in S_x} (e(g,g)^{srq_{parent(z)}(index(z))})^{\delta_{i, S'_x}(0)}$$

$$= \prod_{z \in S_x} e(g,g)^{srq_x(i)\delta_{i, S'_x}(0)}$$

$$= e(g,g)^{srq_x(0)}$$

$$= e(g,g)^{sry}$$

Thus the decryptor can decrypt the ciphertext by compute $M = \frac{E'}{F_x}$.

5. Recover(E, RK): On input the ciphertext and the recovering key, this algorithm computes

$$F = (E'')^r = e(g,g)^{sry}$$

Thus the sender can recover the ciphertext by compute $M = \frac{E'}{F}$.

3.3 Security Proof

Definition 2. *Assume* $(\mathbb{G}_1, \mathbb{G}_2, e)$ *is a bilinear group,* g *is the generator of* $\mathbb{G}_1$. *On input* (g^a, g^b, g^c, Z), *any probabilistic polynomial time algorithm* $\mathcal{A}$ *cannot distinguish* $Z = e(g,g)^{abc}$ *from a random element in* $\mathbb{G}_1$ *with non-negligible probability, this is the DBDH assumption.*

Theorem 1. *If an adversary can break our scheme in the Attribute-based Selective-Set model, then a simulator can be constructed to solve the DBDH problem with a non-negligible advantage.*

Proof. Suppose there exists a polynomial-time adversary $\mathcal{A}$, that can attack our scheme in the Selective-Set model with non-negligible advantage. We build a simulator $\mathcal{B}$ that can play the vDBDH game with non-negligible advantage. The simulation proceeds as follows:

We first let the challenger set the groups $\mathbb{G}_1$ and $\mathbb{G}_2$ with an efficient bilinear map e and generator g. The challenger flips a fair binary coin μ, outside of $\mathcal{B}$'s view. If $\mu = 0$, the challenger sets $(A, B, C, D, Z) = (g^a, g^b, g^c, e(g,g)^{abc})$; otherwise it sets $(A, B, C, D, Z) = (g^a, g^b, g^c, e(g,g)^z)$ for random a, b, c, z. We assume the universe, U is defined. The challenger randomly choose $d \in Z_p$ as the challenged or targeted sender(encrypter)'s recover key (fixed randomness).

1. Init. The simulator $\mathcal{B}$ runs $\mathcal{A}$. $\mathcal{A}$ chooses the set of attributes γ it wishes to be challenged upon.
2. Setup. The simulator sets the parameter $Y = e(g^a, g^b)$. For all $i \in U$, it sets T_i as follows: if $i \in \gamma$, it chooses a random $r_i \in Z_p$ and sets $T_i = g^{r_i}$ (thus, $t_i = r_i$); otherwise it chooses a random $\beta_i \in Z_p$ and sets $T_i = g^{b\beta_i} = B^{\beta_i}$ (thus, $t_i = b\beta_i$). It then gives the public parameters to $\mathcal{A}$.
3. Phase 1.
   - $\mathcal{A}$ adaptively makes requests for the keys corresponding to any access structures Γ such that the challenge set γ does not satisfy T. Suppose $\mathcal{A}$ makes a request for the secret key for an access structure Γ where $\Gamma(\gamma) = 0$. To generate the secret key, $\mathcal{B}$ needs to assign a polynomial Q_x of degree d_x for every node in the access tree Γ.
     We first define the following two procedures: PolySat and PolyUnsat.
   - (a) PolySat($\Gamma_x, \gamma, \lambda_x$): This procedure sets up the polynomials for the nodes of an access sub-tree with satisfied root node, that is, $\Gamma(\gamma) = 1$. The procedure takes an access tree Γ_x (with root node x) as input along with a set of attributes γ and an integer $\lambda_x \in Z_p$. It first sets up a polynomial q_x of degree d_x for the root node x. It sets $q_x(0) = \lambda_x$ and then sets rest of the points randomly to completely

fix q_x. Now it sets polynomials for each child node x' of x by calling the procedure $PolySat(\Gamma_{x'}, \gamma, q_x(index(x')))$. Notice that in this way, $q_{x'}(0) = q_x(index(x'))$ for each child node x' of x.

(b) $\mathsf{PolyUnsat}(\Gamma_x, \gamma, g^{\lambda_x})$: This procedure sets up the polynomials for the nodes of an access tree with unsatisfied root node, that is, $\Gamma_x(\gamma) = 0$. The procedure takes an access tree Γ_x (with root node x) as input along with a set of attributes γ and an element $g^{\lambda_x} \in \mathbb{G}_1$ (where $\lambda_x \in Z_p$).

It first defines a polynomial q_x of degree d_x for the root node x such that $q_x(0) = \lambda_x$. Because $\Gamma_x(\gamma) = 0$, no more than d_x children of x are satisfied. Let $h_x \leq d_x$ be the number of satisfied children of x. For each satisfied child x' of x, the procedure chooses a random point $\lambda_{x'} \in Z_p$ and sets $q_x(index(x')) = \lambda_{x'}$. It then fixes the remaining $d_x - h_x$ points of q_x randomly to completely define q_x. Now the algorithm recursively defines polynomials for the rest of the nodes in the tree as follows. For each child node x' of x, the algorithm calls:

- $\mathsf{PolySat}(\Gamma_{x'}, \gamma, q_x(index(x')))$, if x' is a satisfied node. Notice in this case $q_x(index(x'))$ is known.
- $\mathsf{PolyUnsat}(\Gamma_{x'}, \gamma, g^{q_x(index(x'))})$, if x' is not a satisfied node. Notice that only $g^{q_x(index(x'))}$ can be obtained by interpolation as only $g^{q_x(0)}$ is known in this case.

Notice that in this case also, $q_{x'}(0) = q_x(index(x'))$ for each child node x' of x.

To give keys for access structure Γ, simulator first runs $\mathsf{PolyUnsat}(\Gamma, \gamma, A)$ to define a polynomial q_x for each node x of T. Notice that for each leaf node x of T, we know q_x completely if x is satisfied, if x is not satisfied, then at least $g^{q_x(0)}$ is known (in some cases q_x might be known completely). Furthermore, $q_r(0) = a$.

Simulator now defines the final polynomial $Q_x(\cdot) = bq_x(\cdot)$ for each node x of Γ. Notice that this sets $y = Q_r(0) = ab$. The key corresponding to each leaf node is given using its polynomial as follows. Let $i = att(x)$.

$$D_x = \begin{cases} g^{\frac{Q_x(0)}{t_i}} = g^{\frac{bq_x(0)}{r_i}} = B^{\frac{q_x(0)}{r_i}} & if\ \ att(x)\ \ \in \gamma \\ g^{\frac{Q_x(0)}{t_i}} = g^{\frac{bq_x(0)}{b\beta_i}} = g^{\frac{q_x(0)}{\beta_i}} & otherwise \end{cases}$$

Therefore, the simulator is able to construct a private key for the access structure Γ. Furthermore, the distribution of the private key for Γ is identical to that in the original scheme.

- The simulator also simulates the recover oracle by using the fixed randomness d as $m = \frac{E'}{(E'')^d}$.

4. **Challenge.** The adversary $\mathcal{A}$, will submit two challenge messages m_0 and m_1 to the simulator. The simulator flips a fair binary coin ν, and returns an encryption of m_ν. The ciphertext is output as:

$$E = (\gamma, E' = m_\nu Z^d, E'' = Z, \{E_i = ((g^c)^d)^{r_i}\}_{i \in \gamma})$$

If $\nu = 0$ then $Z^d = e(g,g)^{abcd}$. If we let $s = c, r = d$, then we have

$Y^{rs} = e(g,g)^{abcd}$, $E'' = Z = e(g,g)^{abc}$ and $E_i = ((g^c)^d)^{r_i}$. Therefore, the ciphertext is a valid random encryption of message m_ν.

Otherwise, if $\nu = 1$, then $Z = e(g,g)^z$. We then have $E' = m_\nu e(g,g)^z$. Since z is random, E' will be a random element of $\mathbb{G}_2$ from the adversaries view and the message contains no information about m_ν.

5. **Phase 2.** The simulator simulates the key generation algorithm exactly as it did in Phase 1 except the challenge ciphertexts can not be queried to the recover oracle.

6. **Guess.** $\mathcal{A}$ will submit a guess ν' of ν. If $\nu' = \nu$ the simulator will output $\nu' = 0$ to indicate that it was given a DBDH-tuple otherwise it will output $\nu' = 1$ to indicate it was given a random 4-tuple.

Thus if an adversary can break our scheme in the Attribute-based Selective-Set model, then a simulator can be constructed to solve the DBDH problem with a non-negligible advantage.

4 The Eroca System

Our proposed Eroca system can be seen in Fig. 2. In this system, there are three parties which will play different roles: the autonomous vehicles, the vehicle base stations, and the traffic information cloud. The system roughly runs as the following:

1. The system first setup the parameters and generate the attribute sets and the public/secret keys for different parties. Here the whole system run in an attribute based setting. All the users in the system including autonomous vehicles and the vehicle base stations, all need to have their own key policy which can be expressed as the access structure tree. They will obtain their secret keys corresponding to their key policy via the KP-ABE-SR's Key Generation algorithm.

2. The autonomous vehicles or the vehicle base stations, can first collect the closed road's traffic information and encrypt them using block ciphers like AES using block cipher key, then they encrypt the block cipher key using the attributes as the public keys, via the KP-ABE-SR's Encrypt algorithm, finally they outsource the ciphertexts to the cloud.

3. The cloud continuously push the encrypted related traffic information to the vehicle base stations or the autonomous vehicles, the vehicle base stations can act as a location based traffic management center. The vehicles then decrypt the ciphertexts and get the closely related traffic information which can navigate the vehicles's driving.

4. The autonomous vehicles can also retrieve the traffic information ciphertexts from the cloud by using ABE-SR primitive, and then recover the block cipher key from the ciphertexts via the ABE-SR's Decrypt algorithm, finally they can recover the traffic information by using AES's decryption algorithm. Then they can update these traffic information by encrypting the new traffic information. Note these updates can be very often and traditional ABE can not support this dynamically updating.

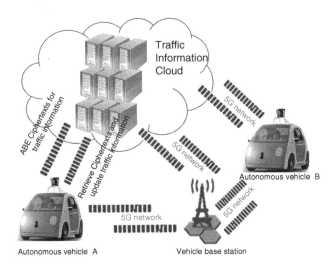

Fig. 2. Our proposed system model

5. The autonomous vehicles can also directly communicate with using ABE-SR to ensure the security and privacy. Also they need often update the closely road's traffic information dynamically, and these information often will be stored in their communicated parters' embedded equipments. In this case, ABE-SR can support the sender autonomous vehicles updating the traffic information more easily compared with traditional ABE.

Here we describe the security objectives of our autonomous vehicle system model and very roughly analysis why the ABE-SR can archive the objects.

1. For autonomous vehicles to achieve privacy and security, their encrypted traffic information need be kept confidential for the adversary, thus the ciphertexts should achieve IND-CPA or IND-CCA security, that is, the ciphertexts should be indistinguishable for the adversary under chosen plaintext attack or chosen ciphertext attack. From the security analysis of our proposed ABE-SR scheme, we can conclude our scheme satisfy this requirement.
2. For autonomous vehicles to easily retrieve and updates the ciphertexts, our ABE-SR scheme provides the recovering and updating function. Thus it is more easily to be use than traditional ABE. Furthermore, the adversary can not get the recovering key, for it is perfectly hidden in the ciphertexts.
3. The cloud or the vehicle base station can not easily get the sensitive traffic information of the autonomous vehicles, for these information has been encrypted by using the block cipher and the ABE-SR scheme.

5   Performance Evaluation

In this section, we evaluate our system's performance roughly. First we compare our system with other existing works, especially on how to integrate retriev-

Table 1. Comparison among EG schemes in the RO model

Protocols	Ciphertexts updatable	Who update	Data owner controlled update	Friendly integrate
[4]	No	-	-	-
[15]	Yes	Cloud	No	Maybe
[6]	Yes	Cloud	No	Maybe
[16]	Yes	Cloud	No	Maybe
[17]	Yes	Cloud	No	Maybe
Eroca	Yes	Data Owner	Yes	Yes

ing and updating data mechanism with attribute based encryption technique smoothly. We can see the comparison results in Table 1.

From this table, we can see our retrieving and updating data mechanism works smoothly with the popular ABE technique, while other mechanisms [12, 18] need to modify or add some other setup to the existing ABE system for achieving updating ciphertexts. In our framework, the data owner implements the data updating operation which means the data owner can control the whole updating process, while other protocols rely on the cloud to update the cipher-text, but we know the cloud can not be trusted, so these protocols can not ensure the cloud implement the updating operation honestly. Chen et al. [18] proposed a new way to efficiently update the large data base with public verifiability, which is not integrated with ABE technique. Lai et al. [19] discussed on how to achieve verifiable computation based on outsourced ciphertexts, but they rely on fully homomorphic encryption which is not very practical and their scheme can not easily integrated ABE.

From the efficiency point, our scheme is also fairly efficient. Compared with the traditional ABE scheme [4], our scheme just adds one more group element to the cipheretxts. And when retrieving the data, the data owner needs only do one modular exponentiation, which can be run on most modern mobile phones quickly. Furthermore, we note our idea can apply to other more efficient ABE schemes and can be more efficient too.

6   Conclusion

In this paper, we propose the notion of attribute based encryption with sender recoverable. Compared with traditional ABE, it can easily achieve retrieve and update the ciphertexts for the encrypter. We propose a concrete KP-ABE-SR scheme and prove its security in the standard model. We also show how to integrate ABE-SR in the autonomous vehicles system which can be a reality in the near future when 5G network is available. We emphasis here that our notion of ABE-SR is just another variant of ABE, it has its own suitable applications and can be used with the traditional ABE schemes smoothly.

Acknowledgment. This work was supported by the Changjiang Scholars and Innovation Research Team in University (Grant NO. IRT 1078), the Key Problem of NFSC-Guangdong Union Foundation (Grant NO. U1135002), the Major Nature Science Foundation of China (Grant NO. 61370078), National High Technology Research and Development Program (863 Program)(No. 2015AA011704), Nature Science Foundation of China (Grant NO. 61103230, 61272492, 61202492, 61402531), Natural Science Foundation of Shaanxi Province (Grant No. 2014JM8300, 2014JQ8358, 2014JQ8307).

References

1. Bethencourt, J., Sahai, A., Waters, B.: Ciphertext-policy attribute-based encryption. In: IEEE Symposium on Security and Privacy (SP07), pp. 321–334 (2007)
2. Piretti, M., Traynor, P., McDaniel, P., Waters, B.: Secure attribute-based systems. In: ACM Conference on Computer and Communications Security (CCS06), pp. 99–112 (2006)
3. Sahai, A., Waters, B.: Fuzzy identity-based encryption. In: Cramer, R. (ed.) EUROCRYPT 2005. LNCS, vol. 3494, pp. 457–473. Springer, Heidelberg (2005)
4. Goyal, V., Omkant, P., Sahai, A., Waters, B.: Attribute-based encryption for fine-grained access control of encrypted data. In: ACM Conference on Computer and Communications Security (CCS06), pp. 89–98 (2006)
5. Waters, B.: Ciphertext-policy attribute-based encryption: an expressive, efficient, and provably secure realization. In: Catalano, D., Fazio, N., Gennaro, R., Nicolosi, A. (eds.) PKC 2011. LNCS, vol. 6571, pp. 53–70. Springer, Heidelberg (2011)
6. Yu, S., Wang, C., Ren, K., Lou, W.: Achieving secure scalable and fine-grained data access control in cloud computing. In: Proceedings of the 29th Conference on Information Communications (INFOCOM 10), pp. 534–542. IEEE Press (2010)
7. Li, J., Chen, X., Li, J., Jia, C., Ma, J., Lou, W.: Fine-grained access control system based on outsourced attribute-based encryption. In: Crampton, J., Jajodia, S., Mayes, K. (eds.) ESORICS 2013. LNCS, vol. 8134, pp. 592–609. Springer, Heidelberg (2013)
8. Li, J., Huang, X., Li, J., Chen, X., Xiang, Y.: Securely utsourcing attribute-based encryption with checkability. IEEE Trans. Parallel Distrib. Syst. **25**, 2201–2210 (2013). doi:10.1109/TPDS.2013.271
9. Ostrovsky, R., Sahai, A., Waters, B.: Attribute-based encryption with non-monotonic access structures. In: ACM Conference on Computer and Communications Security (CCS07), pp. 195–203 (2007)
10. Hur, J., Noh, D.: Attribute-based access control with efficient revocation in data outsourcing systems. IEEE Trans. Parallel Distrib. Syst. **22**(7), 1214–1221 (2011)
11. Li, M., Yu, S., Zheng, Y., Ren, K., Lou, W.: Scalable and secure sharing of personal health records in cloud computing using attribute- based encryption. IEEE Trans. Parallel Distrib. Syst. **24**(1), 131–143 (2013)
12. Yang, K., Jia, X., Ren, K., Zhang, B., Xie, R.: DAC-MACS: effective data acces control for multi-authority cloud storage systems. IEEE Trans. Inf. Forensics Secur. **8**(11), 1790–1801 (2013)
13. Wei, P., Zheng, Y., Wang, X.: Public key encryption for the forgetful. Cryptology ePrint Arch. **2011**, 197 (2011)
14. Wei, P., Zheng, Y.: Efficient public key encryption admitting decryption by sender. In: De Capitani di Vimercati, S., Mitchell, C. (eds.) EuroPKI 2012. LNCS, vol. 7868, pp. 37–52. Springer, Heidelberg (2013)

15. Sahai, A., Seyalioglu, H., Waters, B.: Dynamic credentials and ciphertext delegation for attribute-based encryption. In: Safavi-Naini, R., Canetti, R. (eds.) CRYPTO 2012. LNCS, vol. 7417, pp. 199–217. Springer, Heidelberg (2012)
16. Lee, K., Choi, S.G., Lee, D.H., Park, J.H., Yung, M.: Self-updatable encryption: time constrained access control with hidden attributes and better efficiency. In: Sako, K., Sarkar, P. (eds.) ASIACRYPT 2013, Part I. LNCS, vol. 8269, pp. 235–254. Springer, Heidelberg (2013)
17. Yang, K., Jia, X., Ren, K., Xie, R., Huang, L.: Enabling efficient access control with dynamic policy updating for big data in the cloud. In: INFOCOM14, pp. 2013–2021. IEEE (2014)
18. Chen, X., Li, J., Weng, J., Ma, J., Lou, W.: Verifiable computation over large database with incremental updates. In: Kutyłowski, M., Vaidya, J. (eds.) ICAIS 2014, Part I. LNCS, vol. 8712, pp. 148–162. Springer, Heidelberg (2014)
19. Lai, J., Deng, R.H., Pang, H., Weng, J.: Verifiable computation on outsourced encrypted data. In: Kutyłowski, M., Vaidya, J. (eds.) ICAIS 2014, Part I. LNCS, vol. 8712, pp. 273–291. Springer, Heidelberg (2014)

Coarser-Grained Multi-user Searchable Encryption in Hybrid Cloud

Zheli Liu[1]($\boxtimes$), Chuan Fu[1], Jun Yang[1], Zhusong Liu[2], and Lingling Xu[3]

[1] College of Computer and Control Engineering, Nankai University, Tianjin, China
liuzheli@nankai.edu.cn, {chuanfu,junyang}@mail.nankai.edu.cn
[2] School of Computer Science and Technology,
Guangdong University of Technology, Guangzhou, China
25421944@qq.com
[3] School of Computer Science and Engineering,
South China University of Technology, Guangzhou, China
18826463520@163.com

Abstract. The task of searchable encryption schemes in multi-user setting is to handle the problem of dynamical user injection and revocation with consideration of feasibility. Especially, we have to make sure that user revocation will not cause security problem, such as leakage of secret key. Recently, fine-grained access control using trusted third party is proposed to resolve this issue. However, it increases the management complexity for maintaining massive authentication information of users.

We present a new concept of coarse-grained access control for the first time and use it to construct a multi-user searchable encryption model in hybrid cloud. In our construction, there are two typical schemes, one is broadcast encryption (BE) scheme to simplify access control, the other is a single-user searchable encryption scheme, which supports two-phases operation and is secure when untrustful server colludes with the adversary. Moreover, we implement such a practical scheme using an improved searchable symmetric encryption scheme, and security analysis support our scheme.

Keywords: Multi-user searchable encryption · Identity-based broadcast encryption · Coarse-grained access control · Hybrid cloud

1 Introduction

Searchable encryption (SE) [1] allows "honest but curious" user to search the encrypted files containing some keywords, but reveals as little information as possible to the user. As cloud computing is prevalent, more and more sensitive information is being centralized into clouds, searchable encryption is widely used to protect privacy of users' data.

Many SE schemes [2–5,10,11] have been proposed, including searchable symmetric encryption (SSE) schemes and public key encryption with keyword search (PEKS) schemes. However, these schemes are mostly limited to the single-user

© Springer-Verlag Berlin Heidelberg 2015
N.T. Nguyen et al. (Eds.): Transactions on CCI XIX, LNCS 9380, pp. 140–156, 2015.
DOI: 10.1007/978-3-662-49017-4_9

setting, which allows only one user to perform keyword search and access the encrypted data. We call such schemes single-user searchable encryption (SUSE). In fact, keyword search in multi-user setting is a more common scenario in cloud computing. For example, a user owning a document collection would like to share them with a group of authorized users. We call keyword search in multi-user setting as multi-user searchable encryption (MUSE). Unfortunately, only a few schemes [6–8,12–19] had paid attention and provided help for this scenario.

1.1   Related Works

The first symmetric MUSE scheme was proposed by Curtmola et al. [12] in 2006, which makes use of a SUSE for keyword search and a broadcast encryption (BE) for key management and user management. In this scheme, files owner is broadcaster and group of users with keyword search permission are receivers. The owner shares the key of SUSE to receivers to achieve multi-user searchable encryption. However, sharing key will lead to key exposure when the untrustful server colludes with revoked ones.

Later schemes such as [13,14] focus on how to avoid security problems including key exposure when revoking a user. Dong et al. [13] generate two keys for each authorized user, one is encryption key used by user, the other is proxy key stored in the data server. After receiving the encrypted data of user, the data server will encrypt it again with proxy key. To revoke a user, data server will remove his proxy key. So that if the server can't find the corresponding key for the user, it will abort user's keyword search. In 2008, Bao et al. [14] considered the query on encrypted database and proposed a practical scheme, which uses a user manager (UM) to manage user enrollment and revocation. The scheme [13] divides user's key into two parts, one is for user and the other is for database server, which leads to the two-phase encryption of keywords or trapdoor in both client side and server side.

Recently, other schemes [15–19] were also proposed. Hwang et al. [16] proposed a multi-user encryption scheme with conjunctive keyword search and Wu et al. [15] discussed secure data sharing with public cloud, but they didn't discuss how to add or revoke a user. In 2011, Yang et al. [17] proposed another MUSE scheme for query on encrypted database, which uses a enterprize server to manage users authorization. In 2012, Zhao et al. [18] and Li et al. [19] proposed two multi-user keyword search schemes based on attribute-based encryption (ABE). Both achieve fine-gained access control by using attribute authority to manage users' attribute and trusted third party to verify user's identity.

1.2   Key Problems

Considering a typical scenario of MUSE, in which one user can not only access some encrypted documents from arbitrary other users, but also share his documents to arbitrary group of users. In particular, he can either allow or forbid a user to share his document at any time. Thus, fine-gained access control for each document will be a very complex work. To propose an ideal and practical

MUSE scheme. There are two basic requirements for this: (1) it must handle the problem of dynamical user injection and revocation, especially to make sure that user revocation does not cause security issues like secret key leakage; (2) it should reduce the complexity of user management, and make sure that less or no keyword ciphertext change when adding or removing a user.

We make a conclusion that two key problems of MUSE are key management and access control. Key management can provide more detailed management of encrypt key and query key, and access control can provide the authorized user's access ability to each document. The proposed schemes have provided some good proposals for them.

- About the key management, Curtmola et al. [12] use BE to simplify key management without using trusted third party, but other schemes are not so. In terms of key usage, the SUSE key is shared for authorized group users in [12], which leads to key exposure; two keys are used in Dong et al. [13] and Bao et al. [14], which are used in two-phase encryption of both client side and server side; two symmetric keys of private cloud are also used in Li et al. [19], one is to generate keyword ciphertext and the other is to verify user's identity.
- About the access control, Curtmola et al. [12] shows a powerful method without using trusted third party, by verifying whether the user can decrypt the random secret r encrypted by data owner using BE. To revoke a user, it only needs that data owner randomly selects a new secret r and encrypts it again. Dong et al. [13] and Bao et al. [14] use trusted management server to perform access control. In particular, Zhao et al. [18] and Li et al. [19] achieve the fine-grained access control, which uses attribute authority to maintain each authorized user's access tree and attributes, and whether user can access one file relies on his attributes.

As a result, BE can simplify key management and access control, but it must resolve the problem of key exposure like in [12]. ABE can provide fine-grained access control, however, it always needs a trusted third party to maintain all users' attributes and necessary information about which and whose files can access, so that it will increase the management complexity. An ideal situation is that data owner can freely add or remove a user to access his arbitrary document, and the MUSE scheme can provide a secure mechanism to avoid information leakage.

1.3   Contributions

We proposed a concept of coarse-grained access control for the first time and used it to construct MUSE scheme for data sharing in hybrid cloud. Coarse-grained access control is based on BE, and it can simplify access control by changing the random value r to add or revoke a user. So it is useful for the keyword search in multi-user cryptographic cloud storage.

Moreover, to resolve the problem of information or key leakage when untrusted public cloud colludes with the revoked user, we introduced the concept of SUSE supporting two-phase operation and proposed a practical scheme in hybrid cloud. Under the two-phase operation setting, such schemes will use a private cloud or trusted center to encipher keywords or encipher trapdoors for the second time (it has been encrypted by clients).

2   Preliminaries

In this section, we will review some cryptology concepts used in this paper. In the rest of paper, let U be BE's user space, i.e., the set of all possible user identities, G be a set of users, D be the data to be encrypted, T_r be the trapdoor for keyword search.

2.1   Broadcast Encryption

A broadcast encryption scheme is a tuple of four polynomial-time algorithms $BE = $ *(Setup, Extract, Encrypt, Decrypt)* that work as follows:

- **Setup**(1^k): takes as input a security parameter 1^k and outputs a master key mk.
- **Extract**(mk, id): takes as input a master key mk and a user identier id, then outputs user's key uk_U.
- **Encrypt**(mk, G, m): takes as input a master key mk, a set of users G and a message m, then outputs a ciphertext c.
- **Decrypt**(uk_U, c): takes as input a user key uk_U and a ciphertext c, then outputs either a message m or the failure symbol $\perp$.

Informally, a BE scheme is secure if its ciphertexts leak no useful information about the message to any user not in G.

2.2   Single-User Searchable Encryption

Both SSE and PEKS work in single-user setting, and they can be described as the following tuples $SUSE=$ *(Setup, Encrypt, Trpdr, Search, Decrypt)*:

- **Setup**(1^k): this algorithm is run by the owner to set up the scheme. It takes as input a security parameter 1^k, and outputs the necessary keys.
- **Encrypt**(u, D): this algorithm is run by the owner to encrypt the data D and its keywords. It takes as input the data D, a user u and his necessary keys, then outputs ciphertexts of data and its keywords.
- **Trpdr**(K_w, W): this algorithm is run by a user to generate a trapdoor for a keyword W using key K_w.
- **Search**(T_r): this algorithm is run by the server to perform a search. It takes as input trapdoor T_r, and outputs a set of data or the failure symbol. Before searching, the server may firstly check whether the user has access permission.
- **Decrypt**(u, D): this algorithm is run by the user u to recover the data D.

2.3   Multi-user Searchable Encryption

In this subsection, we will describe a formal definition for multi-user searchable encryption and formalize several security requirements.

A general MUSE scheme is a collection of eight polynomial-time algorithms *MUSE = (Setup, Encrypt, AddUser, RevokeUser, Trpdr, Search, Decrypt, Update)* such that,

- **Setup**(1^k): this algorithm is run by the owner to set up the scheme. It takes as input a security parameter 1^k, and outputs the necessary keys.
- **AddUser**(u): this algorithm is run by the owner to add a user u. It uses owner's secret key to output u's necessary keys. When necessary, the owner maybe update or store some necessary information for access control in trusted third party.
- **RevokeUser**(u): this algorithm is run by the owner to remove a user u. It will delete user u from the set of G. When necessary, the owner maybe delete some necessary information for access control in trusted third party.
- **Encrypt**(D, G): this algorithm is run by the owner to encrypt the data D and its keywords. It takes as input the data D, a set of authorized users G and necessary keys, then outputs ciphertexts of data and keywords.
- **Trpdr**(W): this algorithm is run by a user to generate a trapdoor for a keyword W.
- **Search**(u, T_r): this algorithm is run by the server to perform a search. It takes as input user u and his trapdoor T_r, and outputs a set of data or the failure symbol. Before searching, the server may firstly check whether the user has access permission.
- **Decrypt**(u, D): this algorithm is run by user u to recover the data D.
- **Update**(): this algorithm is run by the owner or some authorized users to update data and its keywords. It is not the necessary algorithm, but it is useful in complex scenarios, such as version control, sharing data in database, etc.

We also formalize several MUSE security requirements, including query privacy and revocability.

Query Privacy. A common security requirement for all searchable encryption schemes is query privacy, which is a security notion on the amount of information leakage to the server regarding user queries. For searchable encryption, server always observes the data access patterns, albeit server is unable to determine the keyword in a query. However, apart from the information that can be acquired via observation and the information derived from it, no other information should be exposed to the server. In the multi-user setting, query privacy must allow user-server collusion.

Revocability. User eviction is an indispensable part of a multi-user application. It is desirable to allow data owner to revoke the search capabilities of users who are deemed no longer appropriate to search the encrypted data. A revoked user might mount attacks on the system or the communication channel in order to perform a search. For instance a replayed query may help a revoked user to

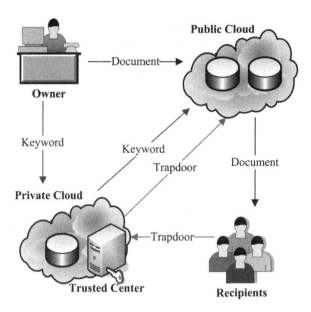

Fig. 1. Basic model of multi-user searchable encryption in hybrid cloud

search the database. An effective approach avoiding this type of attacks can be neutralized by deploying an access control mechanism.

3   System Model

Recently, Bugiel et al. [20] provided an architecture consisting of twin clouds for secure outsourcing of data and arbitrary computations to an untrusted commodity cloud. Based on their twin clouds architecture, we propose a system model for multi-user searchable encryption, which uses private cloud to provide coarse-grained access control and key management, and public cloud to store data. Specially, the private cloud in Bugiel et al.'s work is required to be fully trusted, but the public cloud is only required to be honest.

3.1   Basic Model

A MUSE model should first meet a basic goal: for enciphered data stored in the public cloud, authorized user is able to query over the ciphertexts and get the expectant results, without leaking any information about the queried keywords or the plaintext, even if the adversaries collude with the public cloud. To achieve such goal, system needs to provide access control and key management.

As shown in Fig. 1, there are four different roles in the model, which are *public cloud*, *private cloud*, *owner* and *recipients*.

- *Public cloud* stores data uploaded by users. It is defined as "honest but curious", i.e. it is interested in the users' private data, and could be collude with any user including the revoked users.
- *Private cloud* is allowed to build a trusted center to provide necessary key management and access control mechanism for the whole system. Thus it is required to be fully trusted. In practice, private cloud can be constructed by the enterprises themselves.
- *Owner* is the data provider, who uploads them to the public cloud in order to share them with a group of users.
- *Recipients* are some authorized users, who can access the secret key for deciphering and are allowed to query over enciphered data.

For the data sharing environment in the cloud, it is worth to point out that the owner and recipients are relative. Anyone would share his documents to others, and in this case, he is the data owner. But when to visit some documents shared by others, he will be the recipient. This feature also increases the complexity and difficulty to implement access control.

3.2 Data Flows

As shown in Fig. 1, there are three data flows in the model, including *uploading document*, *generating keyword* and *keyword search*.

- *Uploading document*. To upload document, data owner just only sends the enciphered data to the public cloud.
- *Generating keyword*. When uploading the document, data owner should generate its keywords and store them into the public cloud to support searching over the encrypted data at the same time. To generate the keywords, a "two-phase encryption" is applied: keywords are first enciphered at client and sent to the private cloud, where they are enciphered for the second time and returned to the owner for uploading.
- *Keyword search*. To perform keyword search, recipient should generate the trapdoor for the keyword. Like keyword generation, a "two-phase trapdoor" is also applied: trapdoors are first generated at client and sent to the private cloud, where generated for the second time and returned to the user for searching.

In addition, after getting the encrypted documents by keyword search, recipients need to ask for the corresponding decryption key from the trusted center. The trusted center uses some access control mechanism to verify user's identity, to decide whether or not to send the key based on the user's authority.

3.3 Coarse-Grained Access Control

Although scheme [18,19] with ABE provide fine-grained access control, they always need trusted attribute authority and maintain lots of attributes' information for users and files. Especially for the data sharing environment, the data

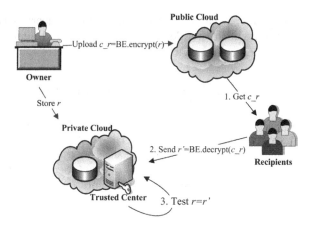

Fig. 2. Coarse-grained access control

owner maybe the recipient at the same time, and the data owner maybe further share the different files to different group of users. These features leads to a large number of information for which or whose files can be accessed to be maintained. In this case, the fine-grained access control using ABE will be not suitable, because more attributes are needed and management complexity is high (Fig. 2).

To provide flexible access control for data sharing, the best approach is to let the data owner to implement access control. Now, we introduce a "**coarse-grained access control**" mechanism, which can make the data owner to freely grant or withdraw the users' authority to access his documents.

The new access control mechanism relies on the BE and a trusted center, and how does it work is like the following:

- *Sharing the random data.* Before sharing the data to a group of users, data owner firstly selects a random value r and store it to the trusted center; then, data owner encrypts the random value r using BE scheme and stores the ciphertext $c\_r = BE.\text{encrypt}(r)$ to the public cloud.
- *Verifying identity.* To verify identity, the user must get the c_r, decrypt it and send the decryption r' to trusted center; the trusted center will test whether r' is equal to r stored in the trusted center.
- *Adding or removing a user.* To add or remove a user, data owner just only needs to select a new random value, stores it and its encryption of BE to the trusted center and public cloud.

The above access control provides a simple way to construct a new MUSE scheme for data sharing under multi-user setting. To construct a practical MUSE scheme, two-phase encryption is also needed, and the coarse-grained access control makes it possible: for preventing curious cloud server from getting information by colluding with the adversary, after verifying the user's identity, the trusted center can also be used to encrypt the keyword or trapdoor from the clients for the second time. Because revoked users can not decrypt the new random value, they can not finish the identity verify again, and hence, the system will be secure even if they collude with the untrusted server.

4 Multi-user Searchable Encryption for Data Sharing

To construct a practical MUSE scheme for data sharing with coarse-grained access control in hybrid cloud, besides the BE, a SUSE scheme supporting two-phase operation is required, which can be described as $SUSETP = (Setup, Encrypt, EncryptSecond, Trpdr, TrpdrSecond, Decrypt, Search)$, where $Encrypt\text{-}Second$ is to further generate keyword ciphertext and $TrpdrSecond$ is to further generate trapdoor. However, there are no such schemes, so we will discuss how to build such a new scheme in Sect. 5.2.

4.1 Our Construction

Now, we provide a description on how to construct MUSE with coarse-grained access control in hybrid cloud, using the selected BE and the selected SUSETP. Let U denote the set of all users and $G \in U$ denote the set of users (currently) authorized to search. The construction is described as follows:

– **Setup**(1^k): the owner generates $K \leftarrow SUSETP.setup(1^k)$ and $mk \leftarrow BE.setup(1^k)$, where K will be shared by users. Furthermore, the owner generates another key K' and stores it to the trusted center for two-phase encryption.
– **Encrypt**(D, G): firstly, the owner samples symmetric encryption key $K_f \leftarrow \{0,1\}^k$ and encrypts file f in data collection D using K_f to get f', computes keyword ciphertexts $c' \leftarrow SUSETP.encrypt(f)$ and performs an interaction to get the final c from trusted center deployed in the private cloud, where $c \leftarrow SUSETP.encryptsecond(c')$ generated by key K'. Then, the owner samples $r \leftarrow \{0,1\}^k$ and computes $st_s \leftarrow BE.encrypt(mk, G, r)$. Finally, the owner uploads (f', c, st_s) to public cloud server and stores the secret r to private cloud.
– **AddUser**(u): the owner generates the private key uk_u for user u with identity id by computing $uk_u \leftarrow BE.extract(mk, id)$, and sends uk_u and K to user.
– **RevokeUser**(u): the owner samples $r \leftarrow \{0,1\}^k$ and computes new $st_s \leftarrow BE.encrypt(mk, G/id, r)$, then replaces the original st_s in public cloud and r in private cloud respectively.
– **Trpdr**(K_u, w): after reviews st_s from public cloud, the user firstly computes $r' \leftarrow BE.decrypt(uk_u, st_s)$ and $T'_r \leftarrow SUSETP.trpdr(K, w)$, then sends r' and T'_r to private cloud. The trusted center in private cloud firstly verifies whether r' is equal to r, if not equal outputs $\bot$, else computes the final $T_r \leftarrow SUSETP.trpdrsecond(T'_r)$ and sends T_r to user.
– **Search**(u, T_r): the public cloud sends $X \leftarrow SUSETP.Search(T_r)$ to user u.
– **Decrypt**(u, f): to obtain the symmetric encryption key K_f, the user sends $r' \leftarrow BE.decrypt(uk_u, st_s)$ to private cloud. After verifying r' is equal to r, the trusted center will send K_f to user u.

Next, we use the scenario of sharing files for multi-user to describe how our construction works.

System Setup. The owner's key is composed of K, K' and mk, where K is shared by authorize users and K' is used for two-phase encryption in trusted center, and both keys are for the SUSETP scheme, but mk is for the BE scheme.

File Uploading. To encrypt a file f in the data collection, the owner first generates a random encryption key K_f and encrypts file using it, then computes keywords by two phase of the operations, respectively $c' \leftarrow SUSETP.encrypt(f)$ in client side and $c \leftarrow SUSETP.encryptsecond(c')$ in the private cloud server. This results in an encrypted file f' and a sequence of keyword ciphertexts c, which will be stored in the public cloud.

To achieve coarse-grained access control, the owner samples a random secret $r \leftarrow \{0,1\}^k$ and encrypts it using BE for a group of users G, and then stores the result st_s into the public cloud, but stores the secret r into the private cloud.

New User Grant. To add a new user u, the owner generates a user key uk_u for the BE scheme and sends it along with the key K of underlying SUSETP scheme to u (thus, the owner acts as the center in a BE scheme).

File Retrieving. To search for a keyword w, user first retrieves the latest state st_s from the public cloud and uses key uk_u to recover r. It generates a single-user trapdoor by computing $T'_r \leftarrow SUSETP.trpdr(K, w)$ and sends (r, T'_r) to the private cloud.

The trusted center in private cloud firstly verifies whether r' is equal to r, if not equal outputs $\bot$, else computes the final $T_r \leftarrow SUSETP.trpdrsecond(T'_r)$ and sends T_r to user. Then, after sending T_r to public cloud, the user can obtain encrypted files X, where $X \leftarrow SUSETP.Search(T_r)$.

To obtain the symmetric encryption key K_f, the user u sends $r' \leftarrow BE.decrypt(uk_u, st_s)$ to private cloud. After verifying r' is equivalent to r, the trusted center will send K_f to user u.

User Revocation. To revoke a user, the owner only needs to sample a new $r \leftarrow \{0,1\}^k$ and re-computes st_s, and then replaces the original st_s in public cloud and r in private cloud.

4.2   Security Analysis

To analyze the security of our construction, and in particular show that the construction satisfies the security requirements given in Sect. 2.3, we assume that the private cloud is trusted but public cloud is "honest-but-curious", and allow the public cloud to collude with the adversaries including revoked users. Especially public cloud will follow our proposed protocol, but try to find out as much secret information as possible based on its possession. Users would try to access data either within or out of the scopes of their privileges. Moreover, the communication channels involving the public cloud are assumed to be insecure.

Therefore, two kinds of adversaries are considered, that is: (1) external adversaries, especially for the public cloud which aims to extract secret information as much as possible; (2) internal adversaries including revoked users and other unauthorized users who aim to obtain more privileges outside of their scopes.

In the proposed construction, ciphertext of keywords stored in public cloud are twice encrypted by both owner and trusted center in private cloud, and if an authorized user wants to perform a keyword search, he must submit the encrypted trapdoor to the trusted center, which verifies user's access permission and encrypts it again. Moreover, the symmetric key is utilized to encrypt file while it is securely stored in trusted center, and if a user wants to get it, he must be authorized by trusted center. Our construction uses a coarse-grained access control building by BE to authorize a user, and only computes trapdoor or sends symmetric key to the users which are authorized. If the adversaries can successfully get the re-encrypted trapdoor or symmetric key with the invalid identity, shows that the selected BE scheme is not secure. If the adversaries can guess the useful information from the keyword ciphertexts or encrypted files without the computation of trusted center, the selected SUSETP scheme will not be secure. Therefore, if selected schemes are all secure, our construction will satisfy the requirement of **query privacy**.

In the proposed construction, if a user wants to perform a keyword search, he must send the trapdoor generated by himself to the trusted center and get re-encrypted trapdoor from it. Because the revoked users can not decrypt the latest state, they will be failed to authorized by the trusted center, and can not finish the keyword search. As a result, if the selected BE scheme is secure, our construction will satisfy the requirement of **revocability**.

4.3   Comparison

Table 1 shows the comparison between our construction and two typical schemes with fine-grained access control.

Table 1. Comparison about access control information.

Schemes	Trusted center	Information for access control	Spacecomplexity
Zhao et al. [18]	Yes	User's key pair	$O(n)$
Li et al. [19]	Yes	User identity and attribute set	$O(n \times m)$
Our construction	Yes	A random value encrypted with BE	$O(1)$

In the Table 1, Zhao et al. [18] and Li et al. [19] both need the trusted center to store the information of access control:

– In Zhao et al. [18], key pair $< UK, CloudUK >$ are generated for each user, where the key $CloudUK$ is stored in cloud server, and so that the space complexity is $O(n)$, where n is the number of valid users; to verify a user's permission, it will check whether the corresponding key exists in the cloud server; when revoking a user, trusted center will instruct cloud server to delete his corresponding key.

- In Li et al. [19], one table T storing authorized user identity and his attribute set is created by each data owner, and so that the space complexity is $O(n \times m)$, where n is the number of data owners and m is max number of users in each group; to verify a user's permission, private cloud will check whether user's identities in the table T; when revoking a user, private cloud will delete his record in table T.

In our construction, the access control is implemented by data owner, and the space complexity is $O(1)$. So, coarse-grained access control will simply the possession of access control and be useful for cloud storage in multi-user setting.

5 Implementation

To implement data sharing system using proposed MUSE scheme with coarse-grained access control in the cloud, an enterprise might use a public cloud service, such as Amazon S3 for achieved data but continue to maintain in-house storage for operational customer data. Alternatively, the trusted private cloud could be a cluster of visualized cryptographic co-processors, which are offered as a service by a third party and which provide the necessary hardware-based security features to implement a remote execution environment trusted by the users. In practice, the internal server deployed by enterprise itself can also be considered as private cloud. To enhance the data security, we require that the connections between users and private cloud are protected by a secure channel (e.g. via SSL/TLS), but the connections between users and public cloud are all open.

5.1 Access Control Implementation

The trusted center can use some database product (such as SQL server, Mysql, etc.) to store the necessary information for access control. When a user wants to share his documents to a group of recipients, he can store an entry as $< ID,$ $OwnerID, RandomValue, TwoPhaseKey >$ in the database of trusted center, where ID denotes the primary key, $OwnerID$ denotes the identity of data owner, $RandomValue$ denotes the selected random value for access control and $TwoPhaseKey$ denotes the key for two-phase operation.

When a recipient wants to verify his identity for getting shared documents form data owner uID, he must submit the decrypted random value r' and uID to the trusted center. On the reception of this request, the trusted center will execute a SQL query to find which record can satisfy both $RandomValue$ is r' and $OwnerID$ is uID. If no record is found, the trusted center will abort the interaction. To avoid malicious dictionary attack, the trusted center will not response for a user who continuously fails to verify his identity.

Except access control mechanism, the following two schemes are necessary: (1) an efficient BE scheme. Some BE schemes have been proposed, such as [22,23].

We adopt Phan's scheme to implement our system, because it has constant-size secret keys and ciphertexts, and has short public key; (2) a SUSE scheme supporting two-phase operation, however, there are no such schemes, so that we discuss how to build one now.

5.2 SUSE Scheme Supporting Two-Phase Operation

Basic Idea. Pseudo random permutation (PRP) security is basic security requirement for most of symmetric ciphers, thus, SSE schemes always use some PRPs in the phase of encrypting keywords and generating trapdoor. Thus, there are two ways to build a SUSE scheme supporting two-phase operation:

1. Making a PRP of the original SSE scheme to run in the trusted center. The client firstly generates a part of ciphertext, when to encrypt it using the original PRP, the client should require trusted center to do it by running this PRP, and finally, the client continues to generate the final keyword ciphertext.
2. Selecting a new PRP to run in the trusted center. When generating keyword ciphertext, after encrypting by PRP of the original SSE scheme, the client should send the middle result to the trusted center to encrypt it again using the selected new PRP, and finally, the client continues to generate the final keyword ciphertext.

Practical Schemes. There are three typical SSE schemes, i.e., SWP scheme [1], Z-IDX scheme [21] and SSE-1 scheme [12]. SWP scheme is constructed by sequential scanning strategy, which generates all keywords and encrypted them one by one; later two schemes are constructed by building index strategy, which includes two sub-procedures of building index and encrypt file. The strategy of building index can obviously improve the efficiency of keyword search. Based on the basic idea, all three schemes can be improved to SSE schemes supporting two-phase encryption.

As an example, we can directly construct such a new scheme by improving the SWP scheme. We firstly introduce how SWP scheme works. Three algorithms are used in SWP scheme: (1) a pseudorandom generator G, which can generate a sequence of pseudorandom values; (2) a pseudo random function (PRF) $F : X \rightarrow Y$, which can map an element in X into another element in Y; (3) a pseudo random permutation E, i.e., a block cipher, which is used to encrypt the keyword W. As shown in Fig. 3, for the i-th keyword W_i to be encrypted, where $|W_i| = n$, the SWP scheme works as following:

1. *Encrypting keyword.* Firstly, SWP scheme encrypts W_i using E with key K, and gets $E_i \leftarrow E(K, W_i)$. Secondly, it generates a pseudorandom value S_i using G (in practice, the S_i is produced in advance), and sets $T_i = S_i||F(S_i)$, where $A||B$ denotes their concatenation. Finally, SWP scheme outputs $C_i = E_i \oplus T_i$.
2. *Generating trapdoor.* If the user wants to search the word W, he can generate the trapdoor as $T_r = E(K, W)$ and sends it to the cloud server.

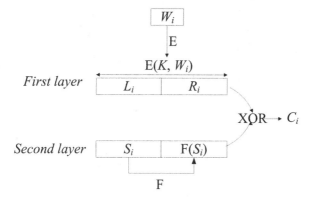

Fig. 3. SWP scheme

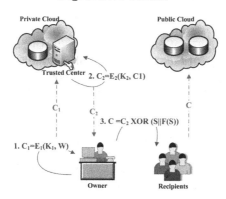

Fig. 4. SUSETP scheme from SWP scheme.

3. *Testing.* The cloud server can search for W in the ciphertext by checking whether $C_i \oplus T_r$ is of the form $s||F(s)$ for some s.

We can see that SWP scheme uses E to encrypt the keyword and generate the trapdoor, and thus we can build a new SUSETP scheme using following methods:

1. *Using new PRP.* As shown in Fig. 4, a new PRP E_2 is used to encrypt the keyword again in the trusted center. When to generate trapdoor, the client also sends the trapdoor to the trusted center to be encrypted by E_2 again.
2. *Using original PRP.* As shown in Fig. 5, the original PRP E is moved to run in the trusted center. When to generate trapdoor, the client also sends the keyword to the trusted center to be encrypted by E.

However, SWP scheme is easy to suffer the frequency attack, that is to say, the cloud storage server can perform attack by analysing the keyword ciphertext frequency in the files. Moreover, SWP scheme is inefficient because it will scan the whole encrypted file for each keyword search.

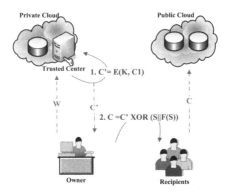

Fig. 5. SUSETP scheme from SWP scheme.

To provide efficient keyword search, SSE-1 scheme uses an array A to store the list L of files collection containing keyword w, furthermore, it builds a look-up table T to store the position of L. When building the look-up table T, SSE-1 scheme uses two pseudo random functions as follows, (1) PRP π: to generate the storage position in table T of keyword w with key $k2$; (2) PRF f: to encrypt keyword w with key $k3$. For keyword search, SSE-1 scheme will generate the trapdoor $T_w = (\pi_{k2}(w), f_{k3}(w))$, where $\pi_{k2}(w)$ is used to get the storage position in table T of keyword w, $f_{k3}(w)$ is used to compute the storage position in array A of list L of keyword w. As a result, PRP π is used in both keyword encryption and trapdoor generation, and thus, we can build an efficient SUSETP scheme like above two methods from SWP scheme.

Security Analysis. The proposed SUSETP schemes are secure. Considering two kinds of adversaries, that is: (1) unauthorized adversaries, who aim to get the response of trusted center; (2) internal adversaries, who can be authorized by trusted center but want to guess which key is used by trusted center. For the unauthorized adversaries, trusted center only responses the query after verifying user's identity by coarse-grained access control based on BE, so that our schemes will be secure if the selected BE scheme is secure; For the Internal adversaries, our schemes are secure because PRPs of SSE-1 scheme have provable and practical security.

6   Conclusion

For the data sharing environment in the cloud, it is worth to point out that the owner and recipients are relative. Anyone would share his documents to others, and in this case, he is the data owner. But when to visit some documents shared by others, he will be the recipient.

In this paper, we proposed a multi-user searchable encryption architecture with coarse-grained access control in hybrid cloud. Coarse-grained access control will provide efficient access control and be useful for multi-user setting in cloud storage. Moreover, we described a practical MUSE scheme using an improved SSE-1 scheme. Security analysis shows that our scheme is secure.

Acknowledgment. This work is supported by the National Natural Science Foundation of China (Nos. 60973141 and 61272423), National Key Basic Research Program of China (No. 2013CB834204), and the Specialized Research Fund for the Doctoral Program of Higher Education of China (Nos. 20100031110030 and 20120031120036).

References

1. Song, X., Wagner, D., Perrig, A.: Practical techniques for searches on encrypted data. In: IEEE Symposium on Security and Privacy, pp. 44–55. IEEE Press (2000)
2. Boneh, D., Di Crescenzo, G., Ostrovsky, R., Persiano, G.: Public key encryption with keyword search. In: Cachin, C., Camenisch, J.L. (eds.) EUROCRYPT 2004. LNCS, vol. 3027, pp. 506–522. Springer, Heidelberg (2004)
3. van Liesdonk, P., Sedghi, S., Doumen, J., Hartel, P., Jonker, W.: Computationally efficient searchable symmetric encryption. In: Jonker, W., Petković, M. (eds.) SDM 2010. LNCS, vol. 6358, pp. 87–100. Springer, Heidelberg (2010)
4. Li, J., Wang, Q., Wang, C.: Fuzzy keyword search over encrypted data in cloud computing. In: INFOCOM 2010, pp. 1–5. IEEE Press (2010)
5. Li, J., Man, H.A., Susilo, W., Xie, D.Q., Ren, K.: Attribute-based signature and its applications. In: Proceeding of the 5th ACM Symposium on Information, Computer and Communications Security (ASIACCS 2010), pp. 60–69. ACM (2010)
6. Li, J., Huang, X.Y., Li, J.W., Chen, X.F., Xiang, Y.: Securely outsourcing attribute-based encryption with checkability. IEEE Trans. Parallel Distrib. Syst. **25**(8), 2201–2210 (2014)
7. Li, J., Chen, X.F., Li, M.Q., Li, J.W., Lee, P., Lou, W.J.: Secure deduplication with efficient and reliable convergent key management. IEEE Trans. Parallel Distrib. Syst. **25**(6), 1615–1625 (2014)
8. Li, J., Kim, K.: Hidden attribute-based signatures without anonymity revocation. Inf. Sci. **180**(9), 1681–1689 (2010). Elsevier
9. Li, J., Wang, Q., Wang, C., Ren, K.: Enhancing attribute-based encryption with attribute hierarchy. Mob. Netw. Appl. (MONET) **16**(5), 553–561 (2011). Springer
10. Bösch, C., Brinkman, R., Hartel, P., Jonker, W.: Conjunctive wildcard search over encrypted data. In: Jonker, W., Petković, M. (eds.) SDM 2011. LNCS, vol. 6933, pp. 114–127. Springer, Heidelberg (2011)
11. Zhao, Y., Chen, X.F., Ma, H., et al.: A new trapdoor-indistinguishable public key encryption with keyword search. J. Wirel. Mob. Netw. **3**, 72–81 (2012)
12. Curtmola, R., Garay, J., Kamara, S., Ostrovsky, R.: Searchable symmetric encryption: improved definitions and efficient constructions. In: Proceedings of the 13th ACM Conference on Computer and Communications Security, pp. 79–88. ACM Press (2006)
13. Dong, C., Russello, G., Dulay, N.: Shared and searchable encrypted data for untrusted servers. J. Comput. Secur. **19**, 367–397 (2011)
14. Bao, F., Deng, R.H., Ding, X., Yang, Y.: Private query on encrypted data in multi-user settings. In: Chen, L., Mu, Y., Susilo, W. (eds.) ISPEC 2008. LNCS, vol. 4991, pp. 71–85. Springer, Heidelberg (2008)

15. Wu, X., Xu, L., Zhang, X.: Poster: a certificateless proxy re-encryption scheme for cloud-based data sharing. In: Proceedings of the 18th ACM Conference on Computer and Communications Security, pp. 869–872 (2011)
16. Hwang, Y.-H., Lee, P.J.: Public key encryption with conjunctive keyword search and its extension to a multi-user system. In: Takagi, T., Okamoto, E., Okamoto, T., Okamoto, T. (eds.) Pairing 2007. LNCS, vol. 4575, pp. 2–22. Springer, Heidelberg (2007)
17. Yang, Y.J., Lu, H., Weng, J.: Multi-user private keyword search for cloud computing. In: Cloud Computing Technology and Science (CloudCom), pp. 264–271 (2011)
18. Zhao, F., Nishide, T., Sakurai, K.: Multi-user keyword search scheme for secure data sharing with fine-grained access control. In: Kim, H. (ed.) ICISC 2011. LNCS, vol. 7259, pp. 406–418. Springer, Heidelberg (2012)
19. Li, J., Li, J., Chen, X., Jia, C., Liu, Z.: Efficient keyword search over encrypted data with fine-grained access control in hybrid cloud. In: Xu, L., Bertino, E., Mu, Y. (eds.) NSS 2012. LNCS, vol. 7645, pp. 490–502. Springer, Heidelberg (2012)
20. Bugiel, S., Nurnberger, S., Sadeghi, A., Schneider, T.: Twin clouds: an architecture for secure cloud computing. In: Workshop on Cryptography and Security in Clouds, LNC, vol. 7025, pp. 32–44 (2011)
21. Goh, E.: Secure indexes. Technical report 2003/216, IACR ePrint Cryptography Archive (2003). http://eprint.iacr.org/2003/216
22. Hu, L., Liu, Z.L., Cheng, X.C.: Efficient identity-based broadcast encryption without random oracles. J. Comput. 5(3), 331–336 (2010)
23. Phan, D.-H., Pointcheval, D., Shahandashti, S.F., Strefler, M.: Adaptive CCA broadcast encryption with constant-size secret keys and ciphertexts. In: Susilo, W., Mu, Y., Seberry, J. (eds.) ACISP 2012. LNCS, vol. 7372, pp. 308–321. Springer, Heidelberg (2012)

Quantum Information Splitting Based on Entangled States

Xiaoqing Tan$^{(\boxtimes)}$, Peipei Li, Xiaoqian Zhang, and Zhihong Feng

Department of Mathematics, College of Information Science and Technology,
Jinan University, Guangzhou 510632, People's Republic of China
ttanxq@jnu.edu.cn

Abstract. Two quantum information splitting protocols are proposed, that one is based on Bell states and another is based on cluster states. Two protocols provide two different ways to complete the process of quantum information splitting. The measurement results of Alice represent the secret information in the first protocol which is (n,n) protocol. The secret information is encoded into Pauli operations in the second protocol which is $(2, 2)$ protocol. The original secret information is recovered when all participants are honest cooperation according to the principle of quantum information splitting. Two protocols take full advantages of the entanglement properties of Bell states and cluster states in different basis to check eavesdropping, that are secure against the intercept and resend attack and entangled ancilla particles attack. We also analyse the efficiency of these two protocols.

Keywords: Quantum information splitting · Bell states · Cluster states · Intercept and resend attack · Entangled ancilla particles attack · Efficiency analysis

1 Introduction

Quantum cryptography takes full advantage of quantum mechanical properties to complete quantum cryptographic tasks. An important application of quantum cryptography is breaking various popular public-key encryption and signature schemes such as RSA and ElGamal. Quantum key distribution (i.e. QKD) [1,2] uses quantum mechanics to guarantee secure communication in quantum cryptography. The first QKD [3] is proposed in 1984 and it enables two parties to produce a shared random secret key known only to them, which can then be used to encrypt and decrypt messages. Later, an important property of QKD [3] is to detect the presence of any third party trying to gain knowledge of the key between the two communicating users. With the deepening of the research, researchers found more and more studying direction, such as quantum information splitting (i.e. QIS) [4], quantum secure direct communication (i.e. QSDC) [5], and so on. In this paper, we focus on the quantum information splitting. Quantum information splitting is a quantum cryptographic work aiming to split

© Springer-Verlag Berlin Heidelberg 2015
N.T. Nguyen et al. (Eds.): Transactions on CCI XIX, LNCS 9380, pp. 157–175, 2015.
DOI: 10.1007/978-3-662-49017-4_10

a secret information among a group of parties. A good information-splitting protocol should allow authorized subsets of parties, named the access structure, to faithfully reconstruct the information, while other unauthorized parties, named the adversary structure, cannot obtain any information about the secret. Following the rapid development of quantum cryptography, the extension of information splitting to the quantum regime [6–13] which has received much theoretical attention. Some protocols have practical application, such as money transfer [14] and voting protocols. Recently, a series of protocols for quantum information splitting have been proposed by using cluster states [15–17], graph states [18–22] or single-qubit state [23]. Some of them share the classical information through quantum mechanism system, others share the quantum information directly. These protocols can be divided into two types which are the entangled states and the product states.

Quantum entanglement [6–9,13] plays an important role in QIS, such as Bell states and cluster states. In References [7,9,13], cluster states are used to complete QIS and genuinely entangled five-qubit state and a Bell-state are used in Reference [8]. We introduce the application of Bell states and cluster states in Sect. 2. The QIS is one of the most important information-theoretically secure cryptographic protocols, just as quantum secret sharing, comprises a dealer and n participants who are interconnected by some sets of classical and quantum channels. The secret information is shared by legal subsets of players acting collaboratively. In this paper, we propose a multi-party quantum information splitting (i.e. MQIS) based on entanglement swapping (i.e. ES) of Bell states. Two-particle maximally entangled states, Bell states, are easily made and stored in laboratory. Due to these good properties, Bell states are used in the first QIS. In the second scheme, the four-particle cluster states are used to complete QIS protocol. Cluster states are as a particular instance of graph states, where the underlying graph is a connected subset of a d-dimensional lattice. Cluster states are especially useful in the context of the one-way quantum computer. In quantum information and quantum computing, a cluster state [24–28] is a type of highly entangled state of multiple qubits. In Reference [24], the generation of cluster states from the two-particle case to the multi-particle case is simple and feasible. Cluster state has been shown some particular characters, such as maximally connected and persistency is better than GHZ states. Cluster states are generated in lattices of particles with Ising type interactions. A cluster $\mathcal{C}$ is a connected subset of a d-dimensional lattice, and a cluster state is a pure state of the qubits located on $\mathcal{C}$. They are different from other types of entangled states such as GHZ states or W states because it is more difficult to eliminate quantum entanglement in the case of cluster states.

The structure of this paper is described as follows. We give the preliminary of two protocols for the first MQIS based on entanglement swapping of Bell states and the second QIS based on four-particle cluster states in Sect. 2. MQIS protocol based on entanglement swapping of Bell states is proposed in Sect. 3. QIS protocol based on four-particle cluster states is proposed in Sect. 4. We also give analysis of two protocols in Sect. 5. At last, we give a conclusion in Sect. 6.

2   Preliminary

2.1   For the First MQIS Protocol

Before giving protocol, we introduce four Bell states under different basis

$$|\varphi^+\rangle = \frac{1}{\sqrt{2}}(|00\rangle + |11\rangle) = \frac{1}{\sqrt{2}}(|++\rangle + |--\rangle),$$

$$|\varphi^-\rangle = \frac{1}{\sqrt{2}}(|00\rangle - |11\rangle) = \frac{1}{\sqrt{2}}(|+-\rangle + |-+\rangle),$$

$$|\psi^+\rangle = \frac{1}{\sqrt{2}}(|01\rangle + |10\rangle) = \frac{1}{\sqrt{2}}(|++\rangle - |--\rangle),$$

$$|\psi^-\rangle = \frac{1}{\sqrt{2}}(|01\rangle - |10\rangle) = \frac{1}{\sqrt{2}}(|+-\rangle - |-+\rangle), \tag{1}$$

where $|+\rangle = \frac{1}{\sqrt{2}}(|0\rangle + |1\rangle)$ and $|-\rangle = \frac{1}{\sqrt{2}}(|0\rangle - |1\rangle)$. There is an explicit correspondence after the entanglement swapping of Bell states

$$\begin{aligned}
|\varphi^+\rangle_{ab} \otimes |\varphi^+\rangle_{cd} = \frac{1}{2}[&|\varphi^+\rangle_{ad}|\varphi^+\rangle_{bc} + |\varphi^-\rangle_{ad}|\varphi^-\rangle_{bc} \\
&+ |\psi^+\rangle_{ad}|\psi^+\rangle_{bc} + |\psi^-\rangle_{ad}|\psi^-\rangle_{bc}], \\
|\varphi^-\rangle_{ab} \otimes |\varphi^+\rangle_{cd} = \frac{1}{2}[&|\varphi^+\rangle_{ad}|\varphi^-\rangle_{bc} + |\varphi^-\rangle_{ad}|\varphi^+\rangle_{bc} \\
&+ |\psi^+\rangle_{ad}|\psi^-\rangle_{bc} + |\psi^-\rangle_{ad}|\psi^+\rangle_{bc}], \\
|\psi^+\rangle_{ab} \otimes |\varphi^+\rangle_{cd} = \frac{1}{2}[&|\varphi^+\rangle_{ad}|\psi^+\rangle_{bc} - |\varphi^-\rangle_{ad}|\psi^-\rangle_{bc} \\
&+ |\psi^+\rangle_{ad}|\varphi^+\rangle_{bc} - |\psi^-\rangle_{ad}|\varphi^-\rangle_{bc}], \\
|\psi^-\rangle_{ab} \otimes |\varphi^+\rangle_{cd} = \frac{1}{2}[&-|\varphi^+\rangle_{ad}|\psi^-\rangle_{bc} + |\varphi^-\rangle_{ad}|\psi^+\rangle_{bc} \\
&- |\psi^+\rangle_{ad}|\varphi^-\rangle_{bc} + |\psi^-\rangle_{ad}|\varphi^+\rangle_{bc}]. \tag{2}
\end{aligned}$$

Assume that each of Bell states corresponds to two classical bits, which is shown in Table 1.

Table 1. The relationship of Bell states and classical bits.

Bell states	$|\varphi^+\rangle$	$|\varphi^-\rangle$	$|\psi^+\rangle$	$|\psi^-\rangle$
Classical bits	00	01	10	11

We will propose one MQIS protocol based on entanglement swapping of Bell states [29], which neither requires local unitary operation [30] nor requires alternative measurements [31] of Bell states. Our aim is to construct a (n, n) QIS protocol that a dealer splits the information into n parts and only n players collaborate to reconstruct the secret information, and the scheme should resist well-known attacks. We will give the details in Sect. 3 and security and efficiency analysis in Sect. 5 for this protocol.

2.2   For the Second QIS Protocol

Cluster state is a kind of simple graph state, i.e. the simple chains or grid structure. Cluster is firstly proposed by Briegel and Raussendorf [28] in 2001. In the
same year, they showed that the scheme of quantum computation that consists
entirely of one-qubit measurements on a particular class of entangled states, the
cluster states [25]. In this paper, we only consider the chain cluster states. They
are one-dimensional and n qubits cluster states and two adjacent particles have
interaction. Therefore, they are also called the linear cluster states (Fig. 1).

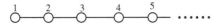

Fig. 1. One-dimensional and n qubits cluster states.

The expression of one-dimensional and n qubits cluster states are

$$|L_N\rangle_{12...N} = (\prod_{1 \leq i \leq N-1} CZ_{i,i+1}|+\rangle^{\otimes N})_{12...N}$$

where $CZ_{i,i+1} = |00\rangle\langle00| + |01\rangle\langle01| + |10\rangle\langle10| - |11\rangle\langle11|$. So, if $N = 2$,

$$|L_2\rangle_{12} = \frac{1}{\sqrt{2}}(|0\rangle|+\rangle + |1\rangle|-\rangle)_{12} \tag{3}$$

Particle 2 can be performed Hadamard transformation in Eq. (3) as

$$|L_2\rangle_{12} = \frac{1}{\sqrt{2}}(|0\rangle|0\rangle + |1\rangle|1\rangle)_{12}$$

where Hadamard transformation is $H = \frac{1}{\sqrt{2}}[(|0\rangle + |1\rangle)\langle0| + (|0\rangle - |1\rangle)\langle1|]$. The
function of H is $H|+\rangle = |0\rangle$, $H|-\rangle = |1\rangle$. Two-particle cluster states are equiva-
lent to Bell states.

If $N = 3$, three-particle cluster states are

$$|L_3\rangle_{123} = \frac{1}{\sqrt{2}}(|+\rangle|0\rangle|+\rangle + |-\rangle|1\rangle|-\rangle)_{123} \tag{4}$$

Particles 1 and 3 can be performed Hadamard transformation in Eq. (4) as

$$|L_3\rangle_{123} = \frac{1}{\sqrt{2}}(|0\rangle|0\rangle|0\rangle + |1\rangle|1\rangle|1\rangle)_{123}$$

So, three-particle cluster states are equivalent to three-particle GHZ states. If
$N = 4$, four-particle cluster states as

$$|L_4\rangle_{1234} = \frac{1}{\sqrt{2}}(|+\rangle|0\rangle|0\rangle|+\rangle + |+\rangle|0\rangle|1\rangle|-\rangle$$
$$+ |-\rangle|1\rangle|0\rangle|+\rangle + |-\rangle|1\rangle|1\rangle|-\rangle)_{1234} \tag{5}$$

Particles 1 and 4 can be performed Hadamard transformation in Eq. (5) as

$$|L_4\rangle_{1234} = \frac{1}{\sqrt{2}}(|0\rangle|0\rangle|0\rangle|0\rangle + |0\rangle|0\rangle|1\rangle|1\rangle$$
$$+ |1\rangle|1\rangle|0\rangle|0\rangle + |1\rangle|1\rangle|1\rangle|1\rangle)_{1234}. \tag{6}$$

Cluster states have two basically properties, i.e. maximum connectedness and better persistency [28]. We also propose the other QIS protocol based on four-particle cluster states in detail in Sect. 4 and analyse the security and efficiency in Sect. 5.

3 MQIS Protocol Based on ES of Bell States

Based on Sect. 2.1, we will propose MQIS protocol based on ES of Bell states. For convenience, we will first present a $(2, 2)$ protocol, then present the $(3, 3)$ protocol. Finally, we present the general (n, n) MQIS protocol.

3.1 $(2, 2)$ Protocol

Now, we propose the $(2, 2)$ protocol. Suppose that there are three parties, Alice, Bob and Charlie. The dealer, Alice, wants to distribute secret information between two parties, Bob and Charlie. The procedure of the $(2, 2)$ protocol is shown in Fig. 2.

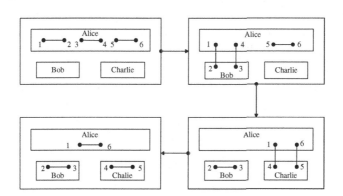

Fig. 2. $(2, 2)$ protocol based on ES of Bell states.

(A.1). Alice prepares three groups enough Bell states, which all are $|\varphi^+\rangle_{12}$, $|\varphi^+\rangle_{34}$, $|\varphi^+\rangle_{56}$. Let S_i is constructed by particles $i(i = 1, 2, ..., 6)$. Every particles sequence S_i prepared by Alice is divided into three parts. The first part is used to check the security of quantum channels. The second part is used to check the honesty of participants. The third part is used to generate the secret of Alice.

The order of sequence $S_i(i=1,2,...,6)$ is disturbed before it is sent in order to ensure the security of quantum channel. She randomly sends S_2 and S_3 to Bob. After receiving S_2 and S_3, Alice tells the order of the particles of S_2 and S_3 to Bob.

(A.2). Alice and Bob begin to check the security of quantum channel between Alice and Bob. Bob randomly chooses the measurement basis (i.e. MB) $\sigma_z = \{|0\rangle, |1\rangle\}$ or $\sigma_x = \{|+\rangle, |-\rangle\}$ to measurement particles 2 in the first part of S_2. Bob tells Alice the MB and measurement results(i.e. MRs). Alice uses the same MB to measure the correlate particle 1 of S_1. Then Alice compares her MRs with Bob's MRs in Eq. (1). If there is no eavesdropping, their outcomes should be completely the same and this process continues. These particles which are used to check eavesdropping are abandoned. Otherwise, QIS is aborted.

(A.3). Alice performs Bell-state measurement(i.e. BSM) on particles 1 and 4. Subsequently, Bob performs BSM on the particles 2 and 3. According to the principle of entanglement swapping, the particles 1, 4 and particles 2, 3 collapse into two Bell states

$$|\varphi^+\rangle_{12} \otimes |\varphi^+\rangle_{34} = \frac{1}{2}[|\varphi^+\rangle_{14}|\varphi^+\rangle_{23} + |\varphi^-\rangle_{14}|\varphi^-\rangle_{23}$$
$$+ |\psi^+\rangle_{14}|\psi^+\rangle_{23} + |\psi^-\rangle_{14}|\psi^-\rangle_{23}]. \tag{7}$$

Alice and Bob complete the transmission of part secret information without the other participants and Bob shares a part of secret information.

(A.4). Alice randomly sends S_4 and S_5 to Charlie. After receiving S_4 and S_5, she tells Charlie the order of particles in S_4 and S_5. Alice and Charlie begin to check the security of quantum channel between Alice and Charlie. The method is same as (A.2). If there is no eavesdropping, this protocol continues. Otherwise, they abort the QIS.

(A.5). Alice performs BSM on particles 1 and 6. Charlie also performs BSM on his particles 4 and 5. The corresponding measurement results are as follows

$$|\varphi^+\rangle_{14} \otimes |\varphi^+\rangle_{56} = \frac{1}{2}[|\varphi^+\rangle_{16}|\varphi^+\rangle_{45} + |\varphi^-\rangle_{16}|\varphi^-\rangle_{45}$$
$$+ |\psi^+\rangle_{16}|\psi^+\rangle_{45} + |\psi^-\rangle_{16}|\psi^-\rangle_{45}],$$
$$|\varphi^-\rangle_{14} \otimes |\varphi^+\rangle_{56} = \frac{1}{2}[|\varphi^+\rangle_{16}|\varphi^-\rangle_{45} + |\varphi^-\rangle_{16}|\varphi^+\rangle_{45}$$
$$+ |\psi^+\rangle_{16}|\psi^-\rangle_{45} + |\psi^-\rangle_{16}|\psi^+\rangle_{45}],$$
$$|\psi^+\rangle_{14} \otimes |\varphi^+\rangle_{56} = \frac{1}{2}[|\varphi^+\rangle_{16}|\psi^+\rangle_{45} - |\varphi^+\rangle_{16}|\psi^-\rangle_{45}$$
$$+ |\psi^+\rangle_{16}|\varphi^+\rangle_{45} - |\psi^+\rangle_{16}|\varphi^-\rangle_{45}],$$
$$|\psi^-\rangle_{14} \otimes |\varphi^+\rangle_{56} = \frac{1}{2}[-|\varphi^+\rangle_{16}|\psi^-\rangle_{45} + |\varphi^-\rangle_{16}|\psi^+\rangle_{45}$$
$$- |\psi^+\rangle_{16}|\varphi^-\rangle_{45} + |\psi^-\rangle_{16}|\varphi^+\rangle_{45}]. \tag{8}$$

(A.6). If Bob and Charlie are honest cooperation, they can deduce the measurement results of Alice in Table 1. Finally, they can obtain the secret information of Alice. For example, if the MRs of Bob and Charlie are $|\psi^-\rangle_{23}$ and $|\varphi^+\rangle_{45}$, they can know that particles 1 and 6 have projected to $|\psi^-\rangle_{16}$ according to Eqs. (7–8). They can obtain the secret information "11" in Table 1.

Note, participants are not all honest. In order to check the honesty of Bob and Charlie, Alice asks that Bob and Charlie publish their MRs in the second part of $S_i(i = 2, 3, 4, 5)$ at the same time. By comparing the MRs of three parties to determine the honesty of Bob and Charlie. If they are honest, this protocol is successfully completed. For example, the MRs of Bob and Charlie are $|\psi^-\rangle_{23}$ and $|\varphi^+\rangle_{45}$, the MRs of Alice are $|\psi^-\rangle_{16}$. Alice know that Bob and Charlie are honest according to the ES of Bell states. Finally, they obtain the useful information in the third part of $S_i(i = 1, 2, 3, 4, 5, 6)$ after these particles which are used to check eavesdropping and honesty are abandoned.

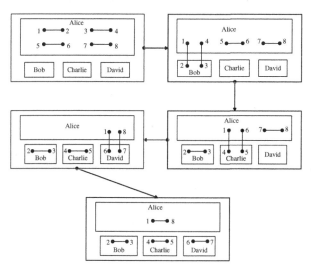

Fig. 3. $(3, 3)$ protocol based on ES of Bell states.

3.2 $(3, 3)$ Protocol

Without loss of generality, we can extend this protocol into a four-party QIS protocol. After analysis, the four-party protocol is secure and reliable. The briefly steps are shown in Fig. 3.

(B.1). Alice prepares four groups enough Bell states, which are all $|\varphi^+\rangle_{12}$, $|\varphi^+\rangle_{34}, |\varphi^+\rangle_{56}, |\varphi^+\rangle_{78}$. Let S_i is constructed by particles $i(i = 1, 2, ..., 8)$. Every particles sequence S_i prepared by Alice is divided into three parts the same as (A.1). The order of sequence $S_i(i=1,2,...,8)$ is disturbed before it is sent. Alice sends S_2 and S_3 to Bob and S_4 and S_5 to Charlie the same as (2,2) QIS protocol.

After Alice finished the BSM on particles 1 and 6, she randomly sends the particles sequences S_6 and S_7 to David. After receiving S_6 and S_7, she tells David the order of particles in S_6 and S_7. Alice and David begin to check eavesdropping the same as (A.2). If there is no eavesdropping, Alice and David perform BSM on their particles.

(B.2). If Bob, Charlie and David are honest cooperation, they can deduce the measurement result of Alice the same as (2,2) QIS and obtain the secret information. For example, if the measurement results of Bob, Charlie and David are respectively $|\varphi^-\rangle_{23}$, $|\varphi^+\rangle_{45}$ and $|\psi^-\rangle_{67}$, they know that the measurement result of Alice is $|\psi^+\rangle_{18}$. Finally, they obtain the secret information "10" in Table 1.

Note, participants are not all honest. In order to check the honesty of Bob, Charlie and David, Alice asks Bob, Charlie and David to public their MRs at the same time the same as $(2,2)$ protocol. By comparing the MRs of three parties to determine the honesty of Bob, Charlie and David. If they are honest, this protocol is successfully completed.

3.3 (n, n) Protocol

So far, we have already presented $(2, 2)$ and $(3, 3)$ protocol based on ES of Bell states. In fact, it is easily generalized to a multiparty case. Suppose that there are $n+1$ parties, Alice and Bob_1, Bob_2, ..., Bob_n. Alice prepares $n+1$ groups enough Bell states, and connects with n participants through one-to-one communication. They can get their measurement results with Bell-state measurement. If Bob_1, Bob_2, ..., and Bob_n are honest cooperation, they can share the secret information each other. The (n, n) QIS protocol is shown in Fig. 4.

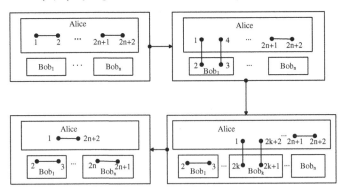

Fig. 4. (n, n) protocol based on ES of Bell states.

(C.1). Alice prepares $n + 1$ groups enough Bell states, which are all $|\varphi^+\rangle_{12}$, $|\varphi^+\rangle_{34}$, ..., $|\varphi^+\rangle_{2n+1,2n+2}$. Let S_i is constructed by particles $i(i = 1, 2, ..., 2k+2)$. Every particles sequence S_i prepared by Alice is divided into three parts the same as (A.1). The order of sequence $S_i(i = 1, 2, ..., 2n + 2)$ is disturbed before it is sent. Alice communicates with Bob_1, the process is the same as (2,2) QIS protocol.

(C.2). Alice randomly sends particles S_{2k} and S_{2k+1} to Bob_k ($k = 2, 3, \cdots, n$). After receiving S_{2k} and S_{2k+1}, she tells Bob_k the order of particles in S_{2k} and S_{2k+1}. Alice and Bob_k begin to check eavesdropping the security. The method is the same as (A.2). If there is no eavesdropping, this protocol continues. Otherwise, they abort the QIS.

(C.3). Alice performs a BSM on particles 1 and $2k + 2$. Bob_k also performs BSM on his particles $2k$ and $2k + 1$. If Bob_1,..., Bob_k,..., and Bob_n are honest cooperation, they can share the secret information (Table 1).

Here, the honesty of participants are checked just as $(2, 2)$ protocol. MQIS protocol is successfully completed.

4   QIS Protocol Based on Four-Particle Cluster States

According to Sect. 2.2, we study the properties of cluster states. We will propose the QIS protocol based on four-particle cluster states. Suppose that there are three participants Alice, Bob and Charlie. The dealer, Alice, wants to distribute secret information between two parties, Bob and Charlie. The procedure of the QIS protocol is shown in Fig. 5.

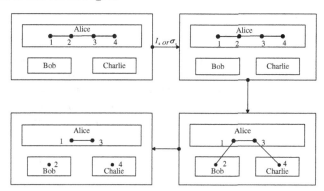

Fig. 5. QIS protocol based on four-particle cluster states

(D.1). Alice prepares two groups enough cluster states $|L_4\rangle_{1234}$ in Eq. (6). One group cluster states D_1 are used to check the security of quantum channel and the other group cluster states D_2 are used to check the honesty and transmit secret information. By simply regrouping terms, the cluster state $|L_4\rangle_{1234}$ of D_1 can be rewritten in the following form

$$|L_4\rangle_{1234} = \frac{1}{2}[|0\rangle|+\rangle|0\rangle|+\rangle + |0\rangle|-\rangle|0\rangle|-\rangle + |1\rangle|+\rangle|1\rangle|+\rangle + |1\rangle|-\rangle|1\rangle|-\rangle]_{1324}$$

$$= \frac{1}{2}[|+\rangle|0\rangle|+\rangle|0\rangle + |-\rangle|0\rangle|-\rangle|0\rangle + |+\rangle|1\rangle|+\rangle|1\rangle + |-\rangle|1\rangle|-\rangle|1\rangle]_{1324} \quad (9)$$

Here, the four states $\{|0+\rangle, |0-\rangle, |1+\rangle, |1-\rangle\}$ form an orthogonal basis MB1 and $\{|+0\rangle, |-0\rangle, |+1\rangle, |-1\rangle\}$ form an orthogonal basis MB2 for the two-qubit Hilbert space. From Eq. (9), we can see that there are four results $|0+0+\rangle$, $|0-0-\rangle$, $|1+1+\rangle$ and $|1-1-\rangle$ with equal probability of $\frac{1}{4}$ if the qubits (1, 3) and (2, 4) in the same basis MB1 or MB2. If one makes a measurement on the particles 1 and 3 in the basis MB1, particles 2 and 4 of the cluster state $|L_4\rangle_{1234}$ will collapse.

(D.2). Firstly, Alice encodes her secret information. Alice selects one of the unitary operation $I = |0\rangle\langle 0| + |1\rangle\langle 1|$ or $\sigma_x = |1\rangle\langle 0| + |0\rangle\langle 1|$ on particles 4 of D_4 according to her secret information.

$$
\begin{aligned}
I|L_4\rangle_{1234} &= \frac{1}{2}[|0\rangle|0\rangle|0\rangle|0\rangle + |0\rangle|0\rangle|1\rangle|1\rangle + |1\rangle|1\rangle|0\rangle|0\rangle + |1\rangle|1\rangle|1\rangle|1\rangle]_{1234} \\
&= \frac{1}{2}[|\varphi^+\rangle_{13}|\varphi^+\rangle_{24} + |\varphi^-\rangle_{13}|\varphi^-\rangle_{24} + |\psi^+\rangle_{13}|\psi^+\rangle_{24} + |\psi^-\rangle_{13}|\psi^-\rangle_{24}] \\
\sigma_x|L_4\rangle_{1234} &= \frac{1}{2}[|0\rangle|0\rangle|0\rangle|1\rangle + |0\rangle|0\rangle|1\rangle|0\rangle + |1\rangle|1\rangle|0\rangle|1\rangle + |1\rangle|1\rangle|1\rangle|0\rangle]_{1234} \\
&= \frac{1}{2}[|\varphi^+\rangle_{13}|\psi^+\rangle_{24} + |\varphi^-\rangle_{13}|\psi^-\rangle_{24} + |\psi^+\rangle_{13}|\varphi^+\rangle_{24} + |\psi^-\rangle_{13}|\varphi^-\rangle_{24}]
\end{aligned}
$$
(10)

After that, the particles 1, 2, 3, 4 of sequence D_1 are randomly inserted into the particles 1, 2, 3 and 4 of sequence D_2, respectively. Let S_i is constructed by particles $i(i = 1, 2, 3, 4)$. Alice reserves sequences S_1 and S_3 and sends particles sequences S_2 and S_4 to Bob and Charlie by quantum channel, respectively.

(D.3). After receiving S_2 and S_4, Alice checks eavesdropping with Bob and Charlie. Alice randomly chooses the measurement basis MB1 or MB2 to measure these particles in D_1. If Alice chooses MB1 to measure her particles, she asks that Bob uses σ_z and Charlie uses σ_x to measure (Eq. (9)). Bob and Charlie publish their MRs at the same time. The other case is similar to MB1. If there is no eavesdropping, these particles in D_1 are abandoned and this protocol continues.

(D.4). Alice performs joint BSM on her particles 1 and 3 of D_2 in S_1 and S_3, Bob and Charlie can choose the MB $\{|0\rangle, |1\rangle\}$ to measure their all particles in S_2 and S_4 of D_2, respectively. After measuring, Alice publishes MRs of particles 1 and 3. If Bob and Charlie are honest cooperation, they can deduce the Pauli operation of Alice. Bob and Charlie successfully obtain the secret information of Alice (Table 2). For example, if the measurement result published by Alice is $|\varphi^+\rangle_{13}$, the corresponding measurement results of Bob and Charlie are all $|0\rangle$. They can deduce that the Pauli operation performed by Alice is I. The classical information 0 is obtained by Bob and Charlie.

Note, the honesty of participants are checked just as (2, 2) QIS protocol. QIS protocol based on four-particle cluster states is successfully completed.

5   Protocol Analysis

Firstly, we present the security analysis of MQIS based on ES of Bell states and QIS based on four-particle cluster states. Secondly, the efficiency analysis is presented.

Table 2. The relationship of participants' MRs and classical information.

Alice's MRs	Bob's MRs	Charlie's MRs	Alice's unitary operation	Classical information
$|\varphi^+\rangle_{13}$	$|0\rangle_2$	$|0\rangle_4$	I	0
		$|1\rangle_4$	σ_x	1
	$|1\rangle_2$	$|1\rangle_4$	I	0
		$|0\rangle_4$	σ_x	1
$|\varphi^-\rangle_{13}$	$|0\rangle_2$	$|0\rangle_4$	I	0
		$|1\rangle_4$	σ_x	1
	$|1\rangle_2$	$|1\rangle_4$	I	0
		$|0\rangle_4$	σ_x	1
$|\psi^+\rangle_{13}$	$|0\rangle_2$	$|1\rangle_4$	I	0
		$|0\rangle_4$	σ_x	1
	$|1\rangle_2$	$|0\rangle_4$	I	0
		$|1\rangle_4$	σ_x	1
$|\psi^-\rangle_{13}$	$|0\rangle_2$	$|1\rangle_4$	I	0
		$|0\rangle_4$	σ_x	1
	$|1\rangle_2$	$|0\rangle_4$	I	0
		$|1\rangle_4$	σ_x	1

5.1   Security Analysis

We will discuss the security of the two protocols. The two protocols are security because the eavesdropper cannot obtain the secret information without being found. In QIS, Alice knows that one of participants, and only one, may be dishonest, and she does not know which is the honest one. So, the honesty of Bob or Charlie is checked. There are two general eavesdropping strategies for Eve, i.e. "intercept and resend" and "entangle and ancilla".

Intercept and Resend Attack

For MQIS based on ES of Bell states and QIS based on cluster states. Before sequences are transmitted, the order of particles are disturbed. Eve is unable to know the correct order. The decoy particles are the signal particles in quantum channel. The decoy particles are in maximally mixed state $\frac{1}{2}(|0\rangle\langle 0| + |1\rangle\langle 1|)$, which means that Eve is unable to distinguish particles. If Eve replaces or measures these particles, this behavior will be found according to Eqs. (1) and (9). The successful probability is $\frac{1}{2}$ in the first scheme and $\frac{1}{4}$ in the second scheme. When the number n of decoy particles is enough big, the detecting probability is $\lim_{n\to\infty} 1 - (\frac{1}{2})^n = \lim_{n\to\infty} 1 - (\frac{1}{4})^n = 1$. Two protocols are secure against this attack.

Entangled Ancilla Particles Attack

We discuss these two protocols respectively.

I. For the First MQIS Protocol. According to the process, the participant Bob_n is the most possible and easy to eavesdrop the secret. So, the security of quantum channel between Alice and Bob_n can represent the security of the whole protocol. As an internal eavesdropper, Bob's purpose is to obtain the measurement result of Alice, and he is permitted to build all devices allowed by the laws of quantum mechanics. Because each particle transmitted in quantum channel is in maximally mixed state, there are no differences among all these particles for Bob_n. Therefore, Bob_n wants to get the secret by entangling ancilla particles.

Let $|\varphi\rangle_{AB_n E_1}$ denotes the state of the composite system including the Bell state $|\chi\rangle_{1,2n}$ and the corresponding ancilla particles E_1, and $|\varphi\rangle_{AB_n E_2}$ denotes the state of the composite system including the Bell state $|\varphi^+\rangle_{2n+1,2n+2}$ and ancilla particles E_2, where A and B_n express the particles belonging to Alice and Bob_n, E_1 and E_2 represent the ancilla particles belonging to Bob_n. If Bob_n obtains no information by measuring the ancilla, it means that $|\varphi\rangle_{AB_n E_1}$ and $|\varphi\rangle_{AB_n E_2}$ must be product of a Bell state and the ancilla.

Without loss of generality, suppose the state of particles 1 and $2n$ is $|\varphi^-\rangle_{1,2n}$. The Schmidt decomposition of $|\varphi\rangle_{AB_n E_1}$ is in the form

$$|\varphi\rangle_{AB_n E_1} = \sum_{i=1}^{4} a_i |\theta_i\rangle_{AB_n} |\varepsilon_i\rangle_{E_1} \tag{11}$$

where $|\theta_i\rangle_{AB_n}$ and $|\varepsilon_i\rangle_{E_1}$ are two sets of orthonormal states, a_i are nonnegative real numbers $(i = 1, 2, 3, 4)$. And $|\theta_i\rangle_{AB_n}$ can be written as the linear combinations of four Bell states

$$\begin{cases} |\theta_1\rangle = b_{11}|\varphi^+\rangle + b_{12}|\varphi^-\rangle + b_{13}|\psi^+\rangle + b_{14}|\psi^-\rangle \\ |\theta_2\rangle = b_{21}|\varphi^+\rangle + b_{22}|\varphi^-\rangle + b_{23}|\psi^+\rangle + b_{24}|\psi^-\rangle \\ |\theta_3\rangle = b_{31}|\varphi^+\rangle + b_{32}|\varphi^-\rangle + b_{33}|\psi^+\rangle + b_{34}|\psi^-\rangle \\ |\theta_4\rangle = b_{41}|\varphi^+\rangle + b_{42}|\varphi^-\rangle + b_{43}|\psi^+\rangle + b_{44}|\psi^-\rangle, \end{cases} \tag{12}$$

where $b_{ij}(i, j = 1, 2, 3, 4)$ are complex numbers. Then $|\varphi\rangle_{AB_n E_1}$ can be written as

$$|\varphi\rangle_{AB_n E_1} = \sum_{i=1}^{4} (a_i b_{i1}|\varphi^+\rangle + a_i b_{i2}|\varphi^-\rangle + a_i b_{i3}|\psi^+\rangle + a_i b_{i4}|\psi^-\rangle)|\varepsilon_i\rangle \tag{13}$$

Similarly, we can get the form of $|\varphi\rangle_{AB_n E_2}$ as

$$|\varphi\rangle_{AB_n E_2} = \sum_{i=1}^{4} (g_i h_{i1}|\varphi^+\rangle + g_i h_{i2}|\varphi^-\rangle + g_i h_{i3}|\psi^+\rangle + g_i h_{i4}|\psi^-\rangle)|\varepsilon_i\rangle \tag{14}$$

where g_i are nonnegative real numbers and h_{ij} are complex numbers$(i = 1, 2, 3, 4)$. For convenience, we define eight vectors as

$$\begin{cases} x_s = (a_1 b_{1s}, a_2 b_{2s}, a_3 b_{3s}, a_4 b_{4s}), s = (1, 2, 3, 4) \\ y_s = (g_1 h_{1s}, g_2 h_{2s}, g_3 h_{3s}, g_4 h_{4s}), s = (1, 2, 3, 4). \end{cases} \tag{15}$$

According to the principle of ES, we can calculate the probability of every possible measurement result. For example, if Alice gets $|\psi^+\rangle_A$ and Bob_n gets $|\varphi^-\rangle_{B_n}$, we can get

$$
|\psi^+\rangle_A|\varphi^-\rangle_{B_n} \otimes \frac{1}{2}\Big[\sum_{p=1}^{4}\sum_{q=1}^{4}(a_p b_{p1} g_q h_{q4} + a_p b_{p2} g_q h_{q3}
$$

$$
+ a_p b_{p3} g_q h_{q2} - a_p b_{p4} g_q h_{q1})|\varepsilon_p\rangle_{E_1}|\varepsilon_q\rangle_{E_2}\Big] \tag{16}
$$

Therefore, the probability is

$$
P(|\psi^+\rangle_A|\varphi^-\rangle_{B_n}) = \frac{1}{4}\sum_{p=1}^{4}\sum_{q=1}^{4}|a_p b_{p1} g_q h_{q4} + a_p b_{p2} g_q h_{q3}
$$

$$
+ a_p b_{p3} g_q h_{q2} - a_p b_{p4} g_q h_{q1}|^2. \tag{17}
$$

Based on Eq. (2), if Bob_n wants to obtain the information without being found, the probability of $|\psi^+\rangle_A|\varphi^-\rangle_{B_n}$ should be equal to 0, i.e. $P(|\psi^+\rangle_A|\varphi^-\rangle_{B_n}) = 0$ according to Eqs. (15) and (17)

$$
x_1^T y_4 + x_2^T y_3 + x_3^T y_2 - x_4^T y_1 = 0, \tag{18}
$$

where x_s^T is the transpose of x_s.

Similarly, the probabilities of $|\psi^-\rangle_A|\varphi^+\rangle_{B_n}$, $|\varphi^-\rangle_A|\psi^+\rangle_{B_n}$, $|\varphi^+\rangle_A|\psi^-\rangle_{B_n}$ should be equal to 0, we can get

$$
\begin{cases}
x_1^T y_4 + x_2^T y_3 - x_3^T y_2 + x_4^T y_1 = 0 \\
-x_1^T y_4 + x_2^T y_3 + x_3^T y_2 + x_4^T y_1 = 0 \\
-x_1^T y_4 + x_2^T y_3 - x_3^T y_2 + x_4^T y_1 = 0.
\end{cases} \tag{19}
$$

From Eqs. (18–19), we can obtain

$$
x_1^T y_4 = x_2^T y_3 = x_3^T y_2 = x_4^T y_1 = 0. \tag{20}
$$

Actually, the probabilities of $|\varphi^+\rangle_A|\psi^+\rangle_{B_n}$, $|\varphi^-\rangle_A|\psi^-\rangle_{B_n}$, $|\psi^+\rangle_A|\varphi^+\rangle_{B_n}$, $|\psi^-\rangle_A|\varphi^-\rangle_{B_n}$ should be equal to 0, then

$$
\begin{cases}
x_1^T y_3 - x_2^T y_4 + x_3^T y_1 + x_4^T y_2 = 0 \\
x_1^T y_3 - x_2^T y_4 - x_3^T y_1 - x_4^T y_2 = 0 \\
x_1^T y_3 + x_2^T y_4 + x_3^T y_1 - x_4^T y_2 = 0 \\
x_1^T y_3 + x_2^T y_4 - x_3^T y_1 + x_4^T y_2 = 0.
\end{cases} \tag{21}
$$

From Eq. (21), we can obtain

$$
x_1^T y_3 = x_2^T y_4 = x_3^T y_1 = x_4^T y_2 = 0. \tag{22}
$$

And the probabilities of $|\varphi^+\rangle_A|\varphi^+\rangle_{B_n}$, $|\varphi^-\rangle_A|\varphi^-\rangle_{B_n}$, $|\psi^+\rangle_A|\psi^+\rangle_{B_n}$, $|\psi^-\rangle_A|\psi^-\rangle_{B_n}$ should be equal to 0, then

$$
\begin{cases}
x_1^T y_1 + x_2^T y_2 + x_3^T y_3 + x_4^T y_4 = 0 \\
x_1^T y_1 - x_2^T y_2 - x_3^T y_3 - x_4^T y_4 = 0 \\
x_1^T y_1 - x_2^T y_2 + x_3^T y_3 + x_4^T y_4 = 0 \\
x_1^T y_1 - x_2^T y_2 - x_3^T y_3 - x_4^T y_4 = 0.
\end{cases} \tag{23}
$$

From Eq. (23), we can obtain

$$x_1^T y_1 = x_2^T y_2 = x_3^T y_3 = x_4^T y_4 = 0. \tag{24}$$

Finally, we can obtain six outcomes by jointing Eqs. (20), (22) and (24)

1. $x_1 = x_2 = x_3 = x_4 = 0$;
2. $y_1 = y_2 = y_3 = y_4 = 0$;
3. $\begin{cases} x_1 = x_2 = x_3 = 0 \\ y_1 = y_2 = y_4 = 0; \end{cases}$
4. $\begin{cases} x_1 = x_2 = x_4 = 0 \\ y_1 = y_2 = y_3 = 0; \end{cases}$
5. $\begin{cases} x_1 = x_3 = x_4 = 0 \\ y_2 = y_3 = y_4 = 0; \end{cases}$
6. $\begin{cases} x_2 = x_3 = x_4 = 0 \\ y_1 = y_3 = y_4 = 0. \end{cases}$

For the first outcome, we have $|\varphi\rangle_{AB_n E_1} = 0$. For the second outcome, we have $|\varphi\rangle_{AB_n E_2} = 0$. The two outcomes are meaningless for our analysis. According to the rest of outcomes, $|\varphi\rangle_{AB_n E_1}$ and $|\varphi\rangle_{AB_n E_2}$ can be written

$$\begin{cases} |\varphi\rangle_{AB_n E_1} = |\psi^-\rangle_{AB_n} \sum_{i=1}^{4} a_i b_{i4}|\varepsilon_i\rangle_{E_1} \\ |\varphi\rangle_{AB_n E_2} = |\psi^+\rangle_{AB_n} \sum_{i=1}^{4} g_i h_{i3}|\varepsilon_i\rangle_{E_2}, \end{cases} \quad \begin{cases} |\varphi\rangle_{AB_n E_1} = |\psi^+\rangle_{AB_n} \sum_{i=1}^{4} a_i b_{i3}|\varepsilon_i\rangle_{E_1} \\ |\varphi\rangle_{AB_n E_2} = |\psi^-\rangle_{AB_n} \sum_{i=1}^{4} g_i h_{i4}|\varepsilon_i\rangle_{E_2}, \end{cases}$$

$$\begin{cases} |\varphi\rangle_{AB_n E_1} = |\varphi^-\rangle_{AB_n} \sum_{i=1}^{4} a_i b_{i2}|\varepsilon_i\rangle_{E_1} \\ |\varphi\rangle_{AB_n E_2} = |\varphi^+\rangle_{AB_n} \sum_{i=1}^{4} g_i h_{i1}|\varepsilon_i\rangle_{E_2}, \end{cases} \quad \begin{cases} |\varphi\rangle_{AB_n E_1} = |\varphi^+\rangle_{AB_n} \sum_{i=1}^{4} a_i b_{i1}|\varepsilon_i\rangle_{E_1} \\ |\varphi\rangle_{AB_n E_2} = |\varphi^-\rangle_{AB_n} \sum_{i=1}^{4} g_i h_{i2}|\varepsilon_i\rangle_{E_2}. \end{cases}$$

We obtain that $|\varphi\rangle_{AB_n E_1}$ and $|\varphi\rangle_{AB_n E_2}$ are the product of one Bell state and the ancilla. It means that Bob_n cannot obtain any information and this protocol can resist this attack.

II. For the Second QIS Protocol. The analysis is similar to the Case.1. Eve introduces the ancilla $|E\rangle$, and performs the unitary operation U on the particles 2, 4, and $|E\rangle$ as

$$\begin{cases} U|00\rangle_{24}|E\rangle = |00\rangle_{24}|\varepsilon_{00}\rangle + |01\rangle_{24}|\varepsilon_{01}\rangle + |10\rangle_{24}|\varepsilon_{02}\rangle + |11\rangle_{24}|\varepsilon_{03}\rangle \\ U|01\rangle_{24}|E\rangle = |00\rangle_{24}|\varepsilon_{10}\rangle + |01\rangle_{24}|\varepsilon_{11}\rangle + |10\rangle_{24}|\varepsilon_{12}\rangle + |11\rangle_{24}|\varepsilon_{13}\rangle \\ U|10\rangle_{24}|E\rangle = |00\rangle_{24}|\varepsilon_{20}\rangle + |01\rangle_{24}|\varepsilon_{21}\rangle + |10\rangle_{24}|\varepsilon_{22}\rangle + |11\rangle_{24}|\varepsilon_{23}\rangle \\ U|11\rangle_{24}|E\rangle = |00\rangle_{24}|\varepsilon_{30}\rangle + |01\rangle_{24}|\varepsilon_{31}\rangle + |10\rangle_{24}|\varepsilon_{32}\rangle + |11\rangle_{24}|\varepsilon_{33}\rangle, \end{cases}$$

where $|\varepsilon_{ij}\rangle, ij \in (0,1,2,3)$ are ancilla particles states. After that, the whole quantum system becomes

$$\begin{aligned} |\zeta\rangle_{1324E} = \frac{1}{2}[&|00\rangle_{13}(|00\rangle_{24}|\varepsilon_{00}\rangle_E + |01\rangle_{24}|\varepsilon_{01}\rangle_E + |10\rangle_{24}|\varepsilon_{02}\rangle_E + |11\rangle_{24}|\varepsilon_{03}\rangle_E) \\ + &|01\rangle_{13}(|00\rangle_{24}|\varepsilon_{10}\rangle_E + |01\rangle_{24}|\varepsilon_{11}\rangle_E + |10\rangle_{24}|\varepsilon_{12}\rangle_E + |11\rangle_{24}|\varepsilon_{13}\rangle_E) \\ + &|10\rangle_{13}(|00\rangle_{24}|\varepsilon_{20}\rangle_E + |01\rangle_{24}|\varepsilon_{21}\rangle_E + |10\rangle_{24}|\varepsilon_{22}\rangle_E + |11\rangle_{24}|\varepsilon_{23}\rangle_E) \\ + &|11\rangle_{13}(|00\rangle_{24}|\varepsilon_{30}\rangle_E + |01\rangle_{24}|\varepsilon_{31}\rangle_E + |10\rangle_{24}|\varepsilon_{32}\rangle_E + |11\rangle_{24}|\varepsilon_{33}\rangle_E)]. \end{aligned}$$

where $|\zeta\rangle_{1324E}$ denotes the state of the composite system including the cluster state $|L_4\rangle_{1234}$ and the corresponding ancilla particles $|E\rangle$. Then two cases will be discussed, in which Alice measures qubits 1 and 3 under different basis.

(1) Alice, Bob and Charlie measure qubits (1,3) and (2,4) in the basis MB1.

$$|\zeta\rangle_{1324E} = \frac{1}{2}[|v_1\rangle + |v_2\rangle + |v_3\rangle + |v_4\rangle]_{1324E}$$

where

$$\begin{aligned}
|v_1\rangle = \frac{1}{2}[&|0+0+\rangle(|\varepsilon_{00}\rangle + |\varepsilon_{01}\rangle + |\varepsilon_{10}\rangle + |\varepsilon_{11}\rangle) \\
&+ |0+0-\rangle(|\varepsilon_{00}\rangle - |\varepsilon_{01}\rangle + |\varepsilon_{10}\rangle - |\varepsilon_{11}\rangle) \\
&+ |0-0+\rangle(|\varepsilon_{00}\rangle + |\varepsilon_{01}\rangle - |\varepsilon_{10}\rangle - |\varepsilon_{11}\rangle) \\
&+ |0-0-\rangle(|\varepsilon_{00}\rangle - |\varepsilon_{01}\rangle - |\varepsilon_{10}\rangle + |\varepsilon_{11}\rangle)],
\end{aligned} \tag{25}$$

$$\begin{aligned}
|v_2\rangle = \frac{1}{2}[&|0+1+\rangle(|\varepsilon_{02}\rangle + |\varepsilon_{03}\rangle + |\varepsilon_{12}\rangle + |\varepsilon_{13}\rangle) \\
&+ |0+1-\rangle(|\varepsilon_{02}\rangle - |\varepsilon_{03}\rangle + |\varepsilon_{12}\rangle - |\varepsilon_{13}\rangle) \\
&+ |0-1+\rangle(|\varepsilon_{02}\rangle + |\varepsilon_{03}\rangle - |\varepsilon_{12}\rangle - |\varepsilon_{13}\rangle) \\
&+ |0-1-\rangle(|\varepsilon_{02}\rangle - |\varepsilon_{03}\rangle - |\varepsilon_{12}\rangle + |\varepsilon_{13}\rangle)],
\end{aligned} \tag{26}$$

$$\begin{aligned}
|v_3\rangle = \frac{1}{2}[&|1+0+\rangle(|\varepsilon_{20}\rangle + |\varepsilon_{21}\rangle + |\varepsilon_{30}\rangle + |\varepsilon_{31}\rangle) \\
&+ |1+0-\rangle(|\varepsilon_{20}\rangle - |\varepsilon_{21}\rangle + |\varepsilon_{30}\rangle - |\varepsilon_{31}\rangle) \\
&+ |1-0+\rangle(|\varepsilon_{20}\rangle + |\varepsilon_{21}\rangle - |\varepsilon_{30}\rangle - |\varepsilon_{31}\rangle) \\
&+ |1-0-\rangle(|\varepsilon_{20}\rangle - |\varepsilon_{21}\rangle - |\varepsilon_{30}\rangle + |\varepsilon_{31}\rangle)],
\end{aligned} \tag{27}$$

$$\begin{aligned}
|v_4\rangle = \frac{1}{2}[&|1+1+\rangle(|\varepsilon_{22}\rangle + |\varepsilon_{23}\rangle + |\varepsilon_{32}\rangle + |\varepsilon_{33}\rangle) \\
&+ |1+1-\rangle(|\varepsilon_{22}\rangle - |\varepsilon_{23}\rangle + |\varepsilon_{32}\rangle - |\varepsilon_{33}\rangle) \\
&+ |1-1+\rangle(|\varepsilon_{22}\rangle + |\varepsilon_{23}\rangle - |\varepsilon_{32}\rangle - |\varepsilon_{33}\rangle) \\
&+ |1-1-\rangle(|\varepsilon_{22}\rangle - |\varepsilon_{23}\rangle - |\varepsilon_{32}\rangle + |\varepsilon_{33}\rangle)].
\end{aligned} \tag{28}$$

According to Eq. (9), these cases $|0+0-\rangle$, $|0-0+\rangle$ of Eq. (25), $|0+1+\rangle$, $|0+1-\rangle$, $|0-1+\rangle$ and $|0-1-\rangle$ of Eq. (26), $|1+0+\rangle$, $|1+0-\rangle$, $|1-0+\rangle$ and $|1-0-\rangle$ of Eq. (27), $|1+1-\rangle$, $|1-1+\rangle$ of Eq. (28) should not appear. In order to avoid introducing error, we obtain

$$\begin{cases}
|\varepsilon_{00}\rangle = |\varepsilon_{11}\rangle & |\varepsilon_{01}\rangle = |\varepsilon_{10}\rangle \\
|\varepsilon_{02}\rangle = |\varepsilon_{03}\rangle = |\varepsilon_{12}\rangle = |\varepsilon_{13}\rangle = 0 \\
|\varepsilon_{20}\rangle = |\varepsilon_{21}\rangle = |\varepsilon_{30}\rangle = |\varepsilon_{31}\rangle = 0 \\
|\varepsilon_{22}\rangle = |\varepsilon_{33}\rangle & |\varepsilon_{23}\rangle = |\varepsilon_{32}\rangle.
\end{cases} \tag{29}$$

(2) When Alice, Bob and Charlie measure qubits (1,3) and (2,4) in the basis MB2, the method is similar to the above. We can obtain

$$\begin{cases} |\varepsilon_{00}\rangle = |\varepsilon_{22}\rangle & |\varepsilon_{02}\rangle = |\varepsilon_{20}\rangle \\ |\varepsilon_{01}\rangle = |\varepsilon_{03}\rangle = |\varepsilon_{21}\rangle = |\varepsilon_{23}\rangle = 0 \\ |\varepsilon_{10}\rangle = |\varepsilon_{12}\rangle = |\varepsilon_{30}\rangle = |\varepsilon_{32}\rangle = 0 \\ |\varepsilon_{11}\rangle = |\varepsilon_{33}\rangle & |\varepsilon_{13}\rangle = |\varepsilon_{31}\rangle. \end{cases} \tag{30}$$

As a result, from Eqs. (29) and (30), we find that the whole system is in the state

$$|\zeta\rangle_{1324E} = |L_4\rangle_{1324} \otimes |\varepsilon_{00}\rangle_E$$

From the above equation, it shows that $|\zeta\rangle_{1324E}$ is a product of a cluster state $|L_4\rangle_{1324}$ and the ancilla E. This implies that it is impossible for Eve to pass the security testing by entangling the ancilla with the cluster state and measuring it without introducing any error. Case.1 and Case.2 show that two protocols are all against this attack.

Cheating Attack of Participants

As we all known, Alice divides her secret information into several parts because there may be one dishonest participant. Suppose Charlie may be dishonest. Therefore, Alice and honest Bob require to check the honesty of Charlie. In order to alone for obtaining Alice's secret information, Charlie tells Bob fake MRs and Bob doesn't know this cheating behaviour. In the end of every protocol, Alice and Bob all check the honesty of Charlie. The process is as follows. After Bob and Charlie cooperate, Bob and Charlie deduce the secret information of Alice. Then Alice asks them to publish their MRs which are used to check honesty at the same time. Bob ensures that the MRs which Charlie tells him is the same as the MRs published by Charlie. According to the entanglement of cluster state and the ES of Bell states, Alice judges whether Charlie is honest. This attack is valid.

Further, every sequence is transmitted only one time and trojan horse attacking can be avoided in two protocols. In a word, two protocols are secure.

5.2   Efficiency Analysis

Now, we analyze efficiency of two protocols. A figure of merit is the total efficiency η defined as [33,34]

$$\eta = \frac{q_u}{q_t},$$

where q_u is the number of transmitted qubits on the quantum channel, and q_t is the total number of transmitted bits on the classical channel. Here the classical bits which are used to check eavesdropping are neglected. We now study the efficiency of our protocols. So the efficiency of our protocols are $\frac{2n}{2n+2n+2n} = \frac{1}{3}$ and $\frac{n}{4n} = \frac{1}{4}$. The efficiency of the HBB [35], Deng et al.'s protocol [36], and Sun et al.'s protocol [37] are shown in Table 3.

Table 3. The efficiency of different protocols.

	Our first MQIS	Our second QIS	HBB	Deng et al.'s	Sun et al.'s
Efficiency	$\frac{1}{3}$	$\frac{1}{4}$	$\frac{1}{12}$	$\frac{1}{6}$	$\frac{1}{6}$

6 Conclusion

To summarize, we have presented two QIS protocols in this paper. The first one is (n, n) QIS protocol based on entanglement swapping of Bell states and the second one is QIS based on four-particle cluster states. The ways of encoding information are different in two protocols. In the first one, the MRs of Alice are secret information. In the second one, Pauli operations are equivalent to the secret information of Alice. The two schemes take advantages of entangled states under different basis to detect eavesdropping. Therefore, two protocols can against the intercept and resend attack, entangled ancilla particles attack. In the end of every protocol, Alice check the honesty of participants and cheating attack of participants is invalid. Particles are transmitted only one time and trojan horse attacking can be avoided. We also analyze the efficiency of two protocols.

Acknowledgments. The research is funded by National Natural Science Foundation of China, under Grant Nos. 61003258, 61472165, and Science and Technology Planning Project of Guangdong Province, China, under Grant No. 2013B010401018, and Natural Science Foundation of Guangdong Province, China, under Grant No. 2014A030310245, and Guangzhou Zhujiang Science and Technology Future Fellow Fund, under Grant No. 2012J2200094.

References

1. Lo, H.K., Curty, M., Tamaki, K.: Secure quantum key distribution. Nat. Photonics **8**, 595–604 (2014)
2. Sasaki, M., et al.: Field test of quantum key distribution in the Tokyo QKD network. Opt. Express **19**, 10387–10409 (2011)
3. Bennett, C.H., Brassard, G.: Quantum cryptography: public key distribution and coin tossing. In: Proceedings of IEEE International Conference on Computers, Systems and Signal Processing, pp. 175–179 (1984)
4. Karlsson, A., Koashi, M., Imoto, N.: Quantum entanglement for secret sharing and secret splitting. Phys. Rev. A **59**, 162–168 (1999)
5. Liu, Z.H., et al.: Quantum secure direct communication with optimal quantum superdense coding by using general four-qubit states. Quant. Inf. Process **12**, 587–599 (2013)
6. Chi, D.P., Choi, J.W., Kim, J.S., Kim, T., Lee, S.: Quantum states for perfectly secure secret sharing. Phys. Rev. A **78**, 012351-1–012351-4 (2008)
7. Nie, Y.Y., Li, Y.H., Liu, J.C., Sang, M.H.: Quantum information splitting of an arbitrary three-qubit state by using two four-qubit cluster states. Quant. Inf. Process **10**, 297–305 (2011)

8. Nie, Y.Y., Li, Y.H., Liu, J.C., Sang, M.H.: Quantum information splitting of an arbitrary three-qubit state by using a genuinely entangled five-qubit state and a Bell-state. Quant. Inf. Process **11**, 563–569 (2012)

9. Li, Y.H., Liu, J.C., Nie, Y.Y.: Quantum information splitting of an arbitrary three-qubit state by using cluster state and Bell-states. Commun. Theor. Phys. **55**, 421–425 (2011)

10. Luo, M.X., Deng, Y.: Quantum splitting an arbitrary three-qubit state with χ-state. Quant. Inf. Process **12**, 773–784 (2013)

11. Li, D.F., Wang, R.J., Zhang, F.L.: Quantum information splitting of a two-qubit Bell state using a four-qubit entangled state. Chin. Phys. C **39**, 043103-1–043103-5 (2015)

12. Kaushik, N., Goutam, P.: Quantum Information splitting using a pair of GHZ states. Quant. Inf. Comput. **15**, 1041–1047 (2015)

13. Sreraman, M., Prasanta, K.P.: Quantum-information splitting using multipartite cluster states. Phys. Rev. A **78**, 062333-1–062333-5 (2008)

14. Richard, C., Daniel, G., Lo, H.K.: How to share a quantum secret. Phys. Rev. Lett. **83**, 648–651 (1999)

15. Lau, H.K., Christian, W.B.: Quantum secret sharing with continuous-variable cluster states. Phys. Rev. A **88**, 042313-1–042313-10 (2013)

16. Inaba, K., Tokunaga, Y., Tamaki, K., Igeta, K., Yamashita, M.: High-fidelity cluster state generation for ultracold atoms in an optical lattice. Phys. Rev. Lett. **112**, 110501-1–110501-5 (2014)

17. Aguilar, G.H., Kolb, T., Cavalcanti, D., Aolita, L., Chaves, R., Walborn, S.P., Ribeiro, P.H.S.: Linear-optical simulation of the cooling of a cluster-state hamiltonian system. Phys. Rev. Lett. **112**, 160501-1–160501-5 (2014)

18. Markham, D., Sanders, B.C.: Graph states for quantum secret sharing. Phys. Rev. A **78**, 042309-1–042309-17 (2008)

19. Keet, A., Fortescue, B., Markham, D., Sanders, B.C.: Qauntum secret with gubit states. Phys. Rev. A **82**, 062315-1–062315-11 (2010)

20. Sarvepalli, P.: Nonthreshold quantum secret-sharing schemes in the graph-state formalism. Phys. Rev. A **86**, 042303-1–042303-7 (2012)

21. Spengler, C., Kraus, B.: Graph-state formalism for mutually unbiased bases. Phys. Rev. A **88**, 052323-1–052323-21 (2013)

22. Qian, Y.J., Shen, Z., He, G.Q., Zeng, G.H.: Quantum-cryptography network via continuous-variable graph states. Phys. Rev. A **86**, 052333-1–052333-8 (2012)

23. Scherpelz, P., Resch, R., Berryrieser, D., Lynn, T.W.: Entanglement-secured single-qubit quantum secret sharing. Phys. Rev. A **84**, 032303-1–032303-8 (2011)

24. Dong, P., Xue, Z.Y., Yang, M., Cao, Z.L.: Generation of cluster states. Phys. Rev. A **73**, 033818-1–033818-6 (2006)

25. Raussendorf, R., Briegel, H.J.: A one-way quantum computer. Phys. Rev. Lett. **86**, 5188–5191 (2001)

26. Schlingemann, D., Werner, R.F.: Quantum error-correcting codes associated with graphs. Phys. Rev. A **65**, 012308-1–012308-8 (2001)

27. Walther, P., et al.: Experimental one-way quantum computing. Nature **434**, 169–176 (2005)

28. Briegel, H.J., Raussendorf, R.: Persistent entanglement in arrays of interacting particles. Phys. Rev. Lett. **86**, 910–913 (2001)

29. Alexander, M.G., Wagenknecht, C., et al.: Multistage entanglement swapping. Phys. Rev. Lett. **101**, 080403-1–080403-4 (2008)

30. Zhang, Z.J., Man, Z.X.: Multiparty quantum secret sharing of classical messages based ob entanglement swapping. Phys. Rev. A **72**, 022303-1–022303-4 (2005)

31. Li, X., Long, G.L., Deng, F.G., Pan, J.W.: Efficient multiparty quantum-secret-sharing schemes. Phys. Rev. A **69**, 052307-1–052307-5 (2004)
32. Quan, D.X., Zhao, N.: New quantum seret sharing protocol based ob entanglement swapping. J. Optoelectron. Laser **22**, 71–74 (2011)
33. Sun, Y., Du, J.Z., Qin, S.J., et al.: Two-way authentication quantum secret sharing. J. Phys. **57**, 4689–4694 (2008)
34. Long, G.L., Liu, X.S.: Theoretically efficient high-capacity quantum-key-distribution scheme. Phys. Rev. A **65**, 032302-1–032302-3 (2002)
35. Hillery, M., Buzek, V., Berthiaume, A.: Quantum secret sharing. Phys. Rev. A **59**, 1829–1834 (1999)
36. Deng, F.G., Zhou, P., Li, X.H., Li, C.Y., Zhou, H.Y.: Efficient multiparty quantum secret sharing with GHZ states. Chin. Phys. Lett. **23**, 1084-1–1087-3 (2006)
37. Sun, Y., Wen, Q.Y., Gao, F., Chen, X.B., Zhu, F.C.: Multiparty quantum secret sharing based on Bell measurement. Opt. Commun. **282**, 3647–3651 (2009)

Zero-Tree Wavelet Algorithm Joint with Huffman Encoding for Image Compression

Wei Zhang[1]([⊠]), Yuejing Zhang[2], and Aiyun Zhan[2]

[1] College of Software Engineering, East China Jiaotong University,
Nanchang 330013, Jiangxi, People's Republic of China
bear_zw@outlook.com
[2] College of Information Engineering, East China Jiaotong University,
Nanchang 330013, Jiangxi, People's Republic of China
zyjecjtu@foxmail.com, zayscj@163.com

Abstract. Embedded Zero-tree Wavelet (EZW) is an effective image encoding algorithm. This paper emphasizes on the principles of EZW improved algorithm and the realization process for algorithm that includes zero-tree structure, wavelet coefficient scanning mode, improving embedding EZQ algorithm flow. Finally, Huffman coding was jointed to encoding.

By carefully analyzing EZW algorithm, we set the edge threshold as a significant coefficient and querying it with maximum value to determine whether it's the zero-tree root or isolated zero. If the maximum is greater than the threshold, then it will be an isolated zero. The improved algorithm will replace the arithmetic coding method with Huffman coding making it more simpler.

Finally, we simulated the improved algorithm in Matlab to validate the result. Our result shows that in comparison with independent EZW algorithm, the improved algorithm not only increases the compression ratio and encoding efficient, but also improves the peak signal to noise ratio of images and make the vision more clear. Hence proved that the improved algorithm is more feasible and effective.

Keywords: EZW · Huffman coding · Joint coding · Coding efficiency

1 Introduction

Today; increasingly, the information has gone deep into every field of human life, most of information acquirement is originated from image, such as Remote Sensing Image, Medical image, weather satellite images, Digital Television etc. To store, process, and transfer those images have become the primary problem in information times, due to colossal amount of content in it. Therefore the image information will occupy a large amount of storage space, that will eventually increase the cost of information acquirement. With the network developed rapidly, large amounts of information redundancy cannot meet people access to information, therefore; the image compression has become an important resolution. Digital image compression is to reduce the significance of the image storage

© Springer-Verlag Berlin Heidelberg 2015
N.T. Nguyen et al. (Eds.): Transactions on CCI XIX, LNCS 9380, pp. 176–185, 2015.
DOI: 10.1007/978-3-662-49017-4_11

space but also can reduce the burden of the transmission channel, in this way the image processing time can be shortened. So, store and transmit images in compressed type is a meaningful solution to resolve the problem.

Embedded Zero tree Wavelet Coding (EZW) is a simple yet very effective image compression algorithm, proposed by Shapiro in 1993. EZW algorithm adapted the amplitude correlation of different scale layers in wavelet domain to predict and sort, which can remove the correlation between pixels, and meanwhile keep the elaborate structure in different resolutions. Therefore, EZW can realize the progressive coding of some important factors and compression effectively [1]. The main scan code flow of EZW is made of P coefficient, N coefficient, T coefficient and Z coefficient; using the two bit fixed-length code to encode. Nevertheless, the symbol in main scan code flow is not based on equal probability distribution. As a result, the compression ratio cannot acquire higher result with the fixed-length code. On the basis of EZW according to joint encoding approach, we encode to the main scan code flow. The experimental results indicate that the compression ratio can be improved by using this algorithm.

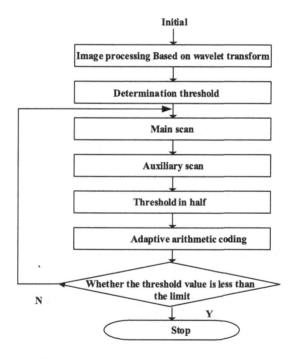

Fig. 1. The flow chart of EZW algorithm

2   Description of EZW Algorithm

EZW is an effective image compression algorithm with Zero-tree [2]. In the existing research, if a wavelet coefficient of low frequency is lower than an amplitude threshold T, the sub ordinate is likely to be lower than T too. With this correlation, the high frequency sub-band coefficient can be predicted by the low frequency sub-band coefficient. The zero-tree encoding of EZW algorithm is a method based on this correlation [3]. As shown in Fig. 1, the steps of EZW algorithm can be performed as follows:

(1) Determine the Amplitude Threshold. The original amplitude threshold is the number no more than the square of absolute value of the coefficient. The equation can be written as follows: $T_0 = 2^{\{\log_2[\max|c_{ij}|]\}}$, where c_{ij} represents for the coefficient of level M wavelet transform, $|c_{i,j}|$ denotes an absolute value of c_{ij}. In $T_i = (1/2)^i T_0$ represents for the amplitude threshold of i+1 times encoding [4,5]. $T_i \geq T_{\min}$ is determined by the expected bit rate. The encoding step will not stop until $T_i = T_{\min}$.

(2) The main scanning. During the scanning process, the coefficients are determined throughout as follows: First; to input wavelet coefficients, judge the absolute value of the wavelet coefficient whether it is greater than the current threshold value or not. If the absolute value of the wavelet coefficient is greater than the current threshold value the coefficient is a positive number in the main scanning table with P-label, whereas; If it is negative, the main scanning table with N-label. Meanwhile, the output symbol coefficient corresponding position P or negative N labeling is set to zero, in order to avoid the main scan again the next time they encode. Secondly, if the absolute value of the wavelet coefficients is less than the current threshold value, it is determined whether the coefficient is a zero root descendant, if it is not to be encoded is skipped. If it is not zero root descendant, it is determined whether the coefficient of variance is greater than the child edge threshold, if the edge is greater than the threshold value, the output is S or G, and the coefficient value is set to zero, encoding is skipped later. If this factor is not the edge information then find the maximum value of the coefficient of the sons of coefficient, if the maximum value is greater than the current threshold value, the point is an isolated zero, otherwise the point of zero roots. While judging a coefficient of zero tree root, it does not factor in all the descendants scanned and marked with O. The main scanning of EZW algorithm is to scan the wavelet coefficient. Following are the principle of scanning parents node:

(3) The sub scanning. The sub scanning is to quantify the important wavelet coefficients of the main scanning output sign bit POS and NEG, auxiliary table formed via expressing each wavelet coefficient value, which is consisted of 0 and 1 [4]. Assuming that the current threshold value is T_i, then the decoder only obtains the important coefficient of amplitude in the interval of $[2T_i, T_i]$ if no further refinement after the main scanning. EZW encoder via the auxiliary scanning, with 0 or 1 of 1 bit to further describe whether the values in the valid amplitude table is in the upper half or the lower half of the interval. 1 represents that the valid values are in the interval $[2T_i, T_i + T_i/2]$ and 0 represents that the valid values are in the interval $[T_i + T_i/2, T_i]$ [6,7] (Fig. 2).

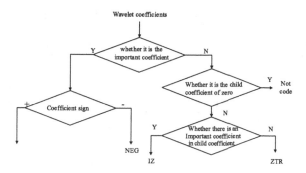

Fig. 2. The scheme of EZW coefficient encoding

(4) Reordering. When the sub-scanning is ended, before scanning the next smaller threshold, in order to improve the decoding precision, and determining the quantized interval of the $i + 1$ times scanning, we need to reorder the data of P, N, S, G according to the principle of a large amplitude at the front of a small principle. Taking example of the threshold Ti, we need to locate the coefficients of amplitude $[3/2T_i, 2T_i]$ coefficients ahead the coefficients of amplitude $[T_i, 3/2T_i]$ [5].

(5) Outputting the encoded information. The information of encoder outputting include the following information: one is to output the information, that is component of the main scanning table, the sub scanning table and current threshold, to encoder; the other is to provide the information, that is component of the current threshold and the reordered significant coefficients, to next scan. Executing the above steps, and updating threshold, then repeating the above procedure.

3   The Analysis of EZW Performance

Table 1 is a 8×8 matrix of wavelet coefficients to describe the methods and results of the two-level EZW coding. The maximum coefficient value is 63. Because 64 is the integer which is the whole power of 2 and bigger than 63. so the first scanning threshold is 64, and the second scan threshold is half, which is 32.

The fine scanning output symbols at the first and second time are as shown in Table 2.

Table 3 only present two scans, wavelet coefficients can be scanned again. Scanning threshold is half in turn, until meet the precision requirement, the scanning can be stopped.

From the performance analysis table, we can suppose that the more encoding times utilized, the lower compression ratio acquired in the same number of decoding. This is because the coding number increases the increase in the number of symbol encoding. Then the number of transmission symbols will increase in the channel. While the compression ratio is the ratio of the amount of data of the original image data and the channel, and thus the compression ratio decreases.

Table 1. The process of wavelet coefficient in the main scanning

63	-34	50	10	7	13	-12	7
-31	23	14	-13	3	4	6	-1
15	14	3	-12	5	-7	3	9
-9	-7	-14	8	4	-2	3	2
-5	9	-1	45	4	6	-2	2
3	0	-3	2	3	-2	0	4
2	-3	6	-4	3	6	3	6
5	11	5	6	0	3	-4	4

(a) Level 3 wavelet transform matrix

P	N	P	T	Z	Z		
Z	T	T	T	Z	Z		
T	Z						
T	T						
		Z	P				
		Z	Z				

(b) The main scanning output symbols at the first time is the search position

0	0	0	10	7	13	-12	7
-31	23	14	-13	3	4	6	-1
15	14	3	-12	5	-7	3	9
-9	-7	-14	8	4	-2	3	2
-5	9	-1	45	4	6	-2	2
3	0	-3	2	3	-2	0	4
2	-3	6	-4	3	6	3	6
5	11	5	6	0	3	-4	4

(c) The main scanning wavelet transform matrix at the second time

Z	T						
N	P						
T	T	T	T				
T	T	T	T				

(d) The main scanning output symbols at the second time is the search position

4   EZW Joint with Huffman Coding Scheme

4.1   The Thought of Joint Coding

In the front EZW encoding process, we know that after the image EZW coding, the transmission code stream of the main scanning in the channel is composed of

Table 2. The fine scanning output symbol flow

The first fine scan		The second fine scan	
The coefficient values	The output symbols	The coefficient values	The output symbols
63	1	63	1
34	0	34	0
50	1	50	0
45	0	45	1
		31	1
		23	0

Table 3. The analysis and comparison about the performance of different codec numbers

The number of coding	The number of decoding	Peak Signal to Noise Ratio PSNR	Compression ratio
4	4	20.3468	20.3848
5	5	19.9024	9.8662

Table 4. The analysis and comparison about the performance of different codec numbers

The number of coding	The number of decoding	The appearance probability of Z	The appearance probability of T	The appearance probability of P	The appearance probability of N
4	4	0.2092	0.6797	0.0753	0.0357
5	5	0.3553	0.4965	0.0849	0.0633

P, N, T, Z. Then every transmission of a symbol is represented by the two binary, Such as "P" in "11", "N" with "10", "T" in "00", "Z" with "01". Hence, the average length of the transmission stream in the channel is 2. In fact, due to the repetition of the output stream, the average code length will be obviously shorter, which can improve the compression ratio, if another joint Huffman coding is done on the output stream and jointly encoded stream is sent to the channel. The problem is that after EZW coding, which can be joint to obtain better performance. Based on the main scanning results of a variety of decoding, we can analyze the probability of the presence of P, N, T, Z. The results are as shown in Table 4.

The probability of Z, T, P, N number appearance only depends on code number. Generally, the probability of occurrence of T and Z are much larger than the probability of P, and N, therefore; it is suitable for joint Huffman coding because the Huffman code is unequal length coding, short codes indicates a high probability, while long codes indicates a low probability and total calculating the average code length is smaller than equal length coding. This results in the achievement of the purpose of compression. Moreover, Huffman encoding is a lossless coding method; it does not affect the image recovery theoretically. EZW joints Huffman coding and decoding processes shown in Fig. 3.

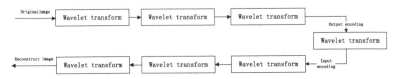

Fig. 3. EZW joint Huffman coding process

4.2   Joint Coding Simulation Results and Performance Analysis

Contrary; the improved EZW coding is introduced in the front, this compression algorithm using Matlab for algorithm simulation, which can be divided into image encoding and image decoding process, including wavelet transform, an improved zero-tree coding. The specific implementation steps of system can be summarized as follows:

(1) The grayscale image can be handled by inputting an original image and reading input file. It can be represented as a two dimensional matrix type that is constituted of image pixels;
(2) Using Haar wavelet for transforming the two-dimensional matrix which will reduce the correlation among coefficients;
(3) Using improved EZW to encode the matrix after transforming, scan with the main scanning first than the sub-scanning.
(4) Encoding the data in the main scanning and the sub-scanning with the Huffman encoding.

The specific simulation process is as follows:
Specific simulation process is as follows:

```
X=imread('flower 512.jpg')
de_x=haardec(X)
[DD,L1,SS,L2]=ezw_encode(de_x,6)
[D,X,sublist]=dominantpass(X,threshold,sublist)
S=subordinatepass(sublist,threshold)
[encode_x,h,sortindex]=huffman(DD)
```

Image reconstruction process is the inverse process of the image encoding:

(1) Using the Huffman decoding the string represented by 0,1 into the symbols of the significant coefficients;
(2) Decoding the symbols for embedded zero-tree-encoded reverse process;
(3) Using Haar wavelet inverse for transforming the decoded data and re-converting the matrix into image.

Table 5. The performance analysis comparison of independent EZW coding and Huffman Joint coding

Code number	Decoding number	Independent EZW coding			EZW joint Huffman coding		
		PSNR	Average code length	Compression ratio	PSNR	Average code length	Compression ratio
4	4	20.3468	2	20.8348	20.3468	1.4313	28.0788
5	5	19.9024	2	9.8662	19.9024	1.6516	11.6768

The specific simulation process is as follows:

```
decode_x=ihuffman(encode_x,h,sortindex)
ezw_decode_x=ezw_decode(de_x,4,decode_x,L1,SS,L2);
re_x=haarrec(ezw_decode_x)
```

The image is similar with encoding, yet makes three-level wavelet decomposition. After the EZW joint Huffman coding, it improves algorithm transmission code from EZW stream to a Huffman code stream.

The reconstructed image effect of different encoding and decoding times between pre-and post of the improved algorithms are shown in Figs. 4 and 5.

The performance analysis comparison of two coding shown in Table 5.

Through comparing both of the above performances after EZW joint Huffman coding; the same code coefficient and PSNR are the compression ratio of joint coding which has been significantly improved.

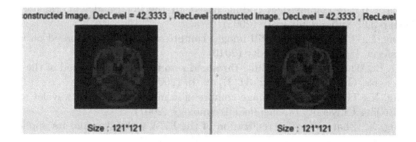

Fig. 4. The reconstructed image of four encoding and four decoding

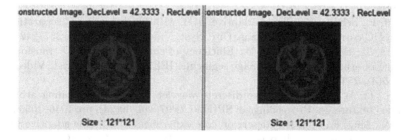

Fig. 5. The reconstructed image of five encoding and five decoding

5   Conclusions

This paper analyzed the existed less of EZW image encoding, and presented an improved algorithm that combine the zero-tree encoding with the Huffman encoding. Through simulation experiments, comparing the objective evaluation parameters of the original algorithm and improved algorithm. The improved algorithm is proved to be not only greatly improving the recovery image subjective visual effect, but also improve the recovery image peak signal to noise ratio. The method joint the EZW with Huffman coding scheme presented in this paper provide a better compression ratio and coding efficiency, after encoding the coefficients of EZW through the Huffman variable length encoding. As a result this reduces the number of bits required for transmission; improve the compression ratio and coding efficiency. Hence, the improved algorithm is effective and feasible.

Acknowledgment. This work is supported by the Jixangxi Province of China Science and Technology Support Program under Grant No.20132BBF60083.

References

1. Zhou, Y.: Embedded zero-tree static image compression method based on wavelet analysis. Comput. Appl. Softw. **21**(8), 119–121 (2004)
2. Cao, Y.: The research of wavelet transform and image compression [D], China University of Geosciences (2005)
3. Wu, R.: Image compression coding based on the 2G Bandelets transform and SPIHT [D], Xidian University (2005)
4. Huang, L.: The research of still images compression technology based on wavelet transform [D], Hunan University (2010)
5. Yang, B., Wang, T., Ye, J.: Multi-threshold zero tree coding method of the images with noise. Optoelectron. Eng. **31**(3), 16–19 (2004)
6. Wang, X.: The research of image compression method based on wavelet analysis [J], Beijing University of Chemical Technology (2007)
7. Zhang, X., Zhang, Z.: The realization of the EZW algorithm and its application. J. Eng. Coll. Armed Police Force **20**(2), 81–84 (2004)
8. Li, S., Li, W., et al.: Shape adaptive wavelet coding. In: Proceedings of the IEEE International Symposium on Circuits and Systems, ISCAS98, vol. 5, pp. 281–284 (2008)
9. Kauff, P., Schuur, K.: Shape-adaptive DCT with block-based DC separation and delta DC correction. IEEE Trans. Circ. Syst. Video Technol. **8**(3), 237–242 (2008)
10. Sikora, T., Bauer, S., Makai, B.: Efficiency of shape-adaptive 2-D transforms for coding of arbitrarily shaped image segments. IEEE Trans. Circ. Syst. Video Technol. **5**(3), 254–258 (2005)
11. Li, S., Li, W.: Shape adaptive discrete wavelet transform for coding arbitrarily shaped texture. In: Proceedings of SPIE VCIP97, vol. 30(24), pp. 1046–1056 (2007)
12. Li, W., Ling, F., Sun, H.: Report on core experiment O3 (Shape adaptive wavelet coding of arbitrarily shaped texture). ISO/IEC JTC/SC29/WG11, MPEG-97-m2385 (2007)

13. Li, S., Li, W., et al.: Shape adaptive vector wavelet coding of arbitrarily shaped texture. ISO/IEC JTC/SC29/WG11, MPEG-96-m1027 (2008)
14. Egger, O.: Region representation using nonlinear techniques withapplications to image and video coding. Ph.D. dissertation, Swiss Federal Institute of Technology (EPFL), Lausanne, Switzerland (2007)
15. Sikora, T., Makai, B.: Shape-adaptive DCT for generic coding of video. IEEE Trans. Circ. Syst. Video Technol. **5**(1), 59–62 (2010)
16. Gilge, M., Engelhardt, T., Mehlan, R.: Coding of arbitrarily shaped image segments based on a generalized orthogonal transform. Sig. Process. Image Commun. **1**(10), 153–180 (2008)
17. Taubman, D.: High performance scalable image compression with EBCOT. IEEE Trans. Image Process. **9**(7), 1158–1170 (2009)
18. Said, A., Pearlman, W.A.: A new fast and efficient image codec based on set partitioning in hierarchical trees. IEEE Trans. Circ. Syst. Video Technol. **6**(3), 243–250 (2005)
19. Shaprio, J.M.: Embedded image coding using zerotree of wavelet coefficients. IEEE Trans. Sig. Process. **41**(12), 3445–3462 (2009)
20. Li, j., Wang, Q., Wang, C., Cao, N., Ren, K., Lou, W.: Fuzzy keyword search over encrypted data in cloud computing. In: proceeding of the 29th IEEE International Conference on Computer Communications (INFOCOM 2010), pp. 441–445. IEEE (2010)
21. Li, J., Kim, K.: Hidden attribute-based signatures without anonymity revocation. Inf. Sci. **180**(9), 1681–1689 (2010). Elsevier
22. Li, J., Wang, Q., Wang, C., Ren, K.: Enhancing attribute-based encryption with attribute hierarchy. Mob. Netw. Appl. (MONET) **16**(5), 553–561 (2011). Springer-Verlag

Author Index

Printed in the United States
By Bookmasters